3000 800062 20696

St. Louis Community College

Florissant Valley Library
St. Louis Community College
3400 Pershall Road
Ferguson, MO 63135-1499
314-513-4514

D1083790

COMPLEXITY IN LANDSCAPE ECOLOGY

Landscape Series

VOLUME 4

Series Editors:

Henri Décamps,
Centre National de la Recherche Scientifique,
Toulouse, France

Bärbel Tress,
Wageningen University,
Wageningen, The Netherlands

Gunther Tress,
Wageningen University,
Wageningen, The Netherlands

Aims & Scope:

The Landscape Series publishes manuscripts approaching landscape from a broad perspective. Landscapes are home and livelihood for people, house historic artefacts, and comprise systems of physical, chemical and biological processes. Landscapes are shaped and governed by human societies, who base their existence on the use of the natural resources. People enjoy the aesthetic qualities of landscapes and their recreational facilities, and design new landscapes. The Landscape Series aims to add new and innovative insights into landscapes. It encourages contributions on theory development as well as applied studies, which may act as best practice. Problem-solving approaches and contributions to planning and management of landscape are most welcome. The Landscape Series wishes to attract outstanding studies from the natural sciences, social sciences, humanities as well as the arts and does especially provide a forum for publications resulting from interdisciplinary and transdisciplinary acting teams. Ideally, the contributions help the application of findings from landscape research to practice, and to feed back again from practice into research.

COMPLEXITY IN LANDSCAPE ECOLOGY

by

David G. Green
Monash University, Melbourne,
VIC, Australia

Nicholas Klomp
Charles Sturt University, Albury,
NSW , Australia

Glyn Rimmington
Wichita State University, Wichita,
KS, U.S.A.

and

Suzanne Sadedin
Monash University, Melbourne,
VIC, Australia

LIBRARY
ST. LOUIS COMMUNITY COLLEGE
AT FLORISSANT VALLEY

 Springer

A C.I.P. Catalogue record for this book is available from the Library of Congress.

ISBN-10 1-4020-4285-X (HB)
ISBN-13 978-1-4020-4285-0 (HB)
ISBN-10 1-4020-4287-6 (e-book)
ISBN-13 978-1-4020-4287-4 (e-book)

Published by Springer,
P.O. Box 17, 3300 AA Dordrecht, The Netherlands.

www.springer.com

Printed on acid-free paper

All Rights Reserved
© 2006, 2007 Springer
No part of this work may be reproduced, stored in a retrieval system, or transmitted
in any form or by any means, electronic, mechanical, photocopying, microfilming, recording
or otherwise, without written permission from the Publisher, with the exception
of any material supplied specifically for the purpose of being entered
and executed on a computer system, for exclusive use by the purchaser of the work.

TABLE OF CONTENTS

Foreword by the series editors

One of the most important aims of Springer's innovative new *Landscape Ecology* series is to bring together the different fields of scientific research which relate to natural and cultural landscapes. Another is to provide an effective forum for dealing with the complexity and range of landscape types that occur, and are studied, globally. In *Complexity and Landscape Ecology*, the third volume of the series, David Green and his co-authors are outstanding in successfully fulfilling both of these aims.

Green and co-authors, Nicholas Klomp, Glyn Rimmington, and Suzanne Sadedin, draw on their wide experience, gained in universities and research institutes across Europe, the United States, Australia, and China, to bring together the fields of ecology, landscapes and complexity theory. With backgrounds spanning areas as diverse as mathematics, computer science, ecology, environmental management and population dynamics, these authors suggest new ways to integrate traditional field ecology within a larger theoretical and management framework. They also demonstrate effectively how the innovative findings of complexity theory can be applied to landscape ecology. Examples include: the network model of complexity; connectivity and connectivity patterns; emergent behaviour and properties; chaos, feedback, modularity and hierarchy; criticality and phase changes.

Given that plants, animals, landscapes and people interact in highly complex and unpredictable ways, it has become necessary, over time, to study the interactions between these different elements, rather than focus only the elements themselves. Green et al. have successfully provided a new framework for further developing these often multidisciplinary studies, and in doing so they make an important contribution to the theoretical advancement of landscape research.

However, the strength of the book is not limited to its theoretical implications, providing, in addition, an effective bridge both between different disciplines and between theory and practice. In view of the need to conserve the world's living resources, the book plays an important role in improving communication between field ecologists, artificial life modellers, biodiversity database developers and conservation managers. At the same time it provides a means for effective communication between academic research and practical landscape management. Written in a vivid and easy to understand language, the authors make a difficult subject easily accessible. As a result, the book will be not only a valuable resource for landscape professionals, but also an interesting introduction for students and the general reader.

Toulouse and Aberdeen, September 2005

Henri Décamps
Bärbel Tress
Gunther Tress

PREFACE

Pick up any typical adventure story, and you are likely to read how the intrepid explorers drop everything and race off into the unknown, where they have hair-raising encounters, see amazing things and eventually win through to their goal. What those stories don't tell you is the boring part - the even longer struggle that the explorers had to endure even to get started on their great adventure. Everyone knows that Columbus discovered the New World. What is often ignored is that he spent long years patiently lobbying at the Spanish court before he even started.

Science tends to be the same way. You read about a great discovery without hearing about the long and often intense struggle that led to it. When we were students, one of our professors used to say that each generation of scientists struggles long and hard to understand some phenomenon. But the next generation accepts that hard-won knowledge as simple and obvious. Newton said that he was able to make his discoveries because he stood on the shoulders of giants. One of these giants was Johannes Kepler, whose three laws of planetary motion were the fruit of an entire lifetime of studying astronomical observations.

It is the same with the science of complexity. As students, we studied science that tended to ignore complexity. It is very satisfying, therefore, to see mainstream ecology beginning to come to terms with complexity. In the course of writing this book, we were surprised to discover just how many ecological studies now adopt the ideas and methods that form an important part of this book. Studies using such techniques as multi-agent simulations and fractal pattern analysis are now commonplace.

The aim of this book is to introduce our readers to the exciting new field of complexity in ecology. Our goal is to provide an easy-to-read introduction. One group of readers we especially hope to serve are people who already have a basic knowledge of, or interest in, ecology, and wish to know what complexity is about. In keeping with these goals, we have tried to keep the book short, rather than have it blow out into a massive tome. Inevitably, we have had to leave out much. This account is in no way intended to be a comprehensive account of the entire field of complexity or landscape ecology. Rather we have chosen to present topics that we hope will provide you with a gentle introduction to this important and exciting area of research.

There is a deliberate trend throughout the book to move from small to large. So we start (Chapter 2) with individuals and even within individuals (in the case of growth and development). At the other extreme, the final chapters deal with large-scale and even global phenomena.

As we explain in the course of the book, simulation models play an important role in studying complexity. We recognise the importance for readers of being able to play these virtual experiments themselves. Therefore we have bundled up many of the models that we describe here as online demonstrations that can be accessed via our Virtual Laboratory web site:

http://www.complexity.org.au/vlab/.

We are indebted to many people who provided material assistance during the writing and production of the book. Tom Chandler and several of his students and colleagues provided images from their virtual reality models for Chapter 9. Our colleagues David Roshier, Gary Luck and David Watson contributed critical comments on several chapters. Joanne Lawrence carried out useful literature surveys during the early stages of writing. Tania Bransden contributed to the editing, indexing and references, and provided a much needed reality check, never letting us get away with lapses into jargon, irrelevance and incomprehensibility! Jeanette Niehus did much of the final formatting, copy editing and proofing of the manuscript. Justine Singh helped with the references and compiled the index. Dr Ann Sadedin and Ruth Cornforth provided useful comments on the manuscript as well as careful proof reading of final drafts.

We are also indebted to the publishers and the series editors for their faith in us and for their encouragement throughout the writing and production of the book. Finally, we are grateful to the Australian Research Council and to the Australian Centre for Complex Systems for funding assistance.

David G. Green
Nick Klomp
Glyn Rimmington
Suzanne Sadedin

Melbourne, August 2005

CHAPTER 1

COMPLEXITY AND ECOLOGY

*Many severe environmental problems have complex causes. This photo shows
a section of the Murray River near Albury in New South Wales, Australia,
recently declared the country's most endangered heritage site in 2002.*

Covering an area of more than 6 million square kilometres, the Amazon Basin dominates northern Brazil and forms a large part of the South American continent. The richness of its biodiversity, and the hostility of its natural environment, mean that even today we can form no clear picture of the Amazon rainforest ecology. Yet even fragmentary glimpses reveal that the Amazon forms an extravagance of nature beyond the wildest imaginings of taxonomists. When biologist T. L. Erwin examined a single species of Amazonian tree, he found 163 unique species of beetle living in its canopy alone (Erwin 1982). Comparing sites seventy kilometres apart, Erwin found only a 1% overlap in the beetle species present (Erwin 1988). Brazil has more species of flowering plants and amphibians than any other country, and ranks in the top four countries on earth for mammals, birds, butterflies and reptiles (Fearnside 1999).

Up until the mid-twentieth century, the Amazon rainforest remained shrouded in mystery, isolated by its sheer size and inaccessibility. But a rapidly growing technological society, with a huge appetite for raw materials, could not leave such a massive resource untouched forever. The forest seemed to promise fertile croplands, and the trees themselves became valuable exports under the chainsaw.

In 1960, the Brasilia-Bellem road opened, making the Amazon accessible for commercial exploitation. The road sliced the rainforest open, bringing swarms of loggers and farmers eager to claim a share of the wealth. During the next two decades, the human population of the region swelled to more than 17 million. Soybean farms and cattle ranches proliferated. In the 1980s, the rate of deforestation across Amazonia exceeded 22,000 square kilometres per year (or about 4 ha per minute). During the late 1990s, the Brazilian government began efforts to reduce the rate of land clearing. However, their attempts were frustrated by the sheer scale of the problem, with illegal operations being responsible for up to 80% of logging in Amazonia. By 1998, over half a million square kilometres of Brazilian rainforest had been cleared (Fearnside 1999), and destruction continues to accelerate. During 2004 alone, over 26,000 square kilometres were destroyed, an area roughly the size of Belgium[1].

The assumption that clearing lush rainforest would yield prime agricultural land proved to be a tragic mistake. Beneath a thin surface layer of rich soil, farmers mostly find infertile wasteland. After a season or two of good crops, the soil is used up, erosion is rife and rainfall declines. Farmers are forced to douse the land with chemical fertilizers, and leave fields fallow for many years, in order to maintain any yield at all. For poverty-stricken smallholders, often faced with an immediate need for cash crops to service World Bank loans, such management techniques are unfeasible. Consequently, millions of displaced farmers move ever deeper into the rainforest, clearing more and more forest to eke out a few extra years' worth of crops.

Ironically, the reason for the land's failure lies in the extraordinary efficiency of the rainforest itself. In the dense, lush ecosystem of the forest, almost nothing is wasted. Nutrients that reach the ground are quickly decomposed and recycled into fast-growing plants. Little is left in the soil: when the trees vanish, so do the raw

---

[1] Brazilian government report, 2005.

materials of life. Even the rain itself depends on the trees: approximately 50% of the Amazon's rainfall is recycled through the forest (Fearnside 1999).

From an ecological viewpoint, the exploitation of the Amazon has been an unmitigated disaster. Logging creates networks of roads, encouraging further migration of farmers into untouched forest. Their slash-and-burn methods break up the forest, gradually turning it into isolated fragments. This fragmentation leads to sharp increases in fires, in hunting, and in soil erosion, as well as invasions of grasses, vines and exotic species. All of these changes spell trouble for native plants and animals.

The story of development in the Amazon Basin is a dramatic example of how simple assumptions about ecological systems can lead to disastrous mistakes in land management. Almost always, problems arise because the complexity of landscapes and ecosystems defeats our efforts to understand them as simple systems of cause and effect. In the case of the Amazon, building a road into the region initiated a cascade of mutually reinforcing processes. The underlying error in this ongoing catastrophe is "cause and effect" thinking: assuming that the forest ecosystem is a direct effect of suitable climate and soil conditions, rather than a complex, dynamic process in itself.

We can see this failure to grasp ecological complexity in early attempts to understand the role of landscapes. When people first began using computer models to study ecosystems, spatial interactions were largely ignored. Local interactions between individuals were assumed to be minor effects that would average out over time and space. Understanding the influence of landscape was therefore seen as easy. To account for a hillside, for instance, all you needed to know was what happened at the top of the hill, at the mid-slope and in the swale. The assumption was that the differences in environmental conditions from place to place were the only factors that influenced the outcome, so accounting for them would tell you all you needed to know about the role of landscapes in ecology.

Unfortunately, the assumption that local effects will average out over time and space is not only incorrect; it is in many cases drastically misleading. Interactions do matter, and local interactions can blow up to have large-scale effects. In ecological systems, many of these interactions are not simple, one-way cause and effect relationships, but complex feedback relationships. Only by explicitly studying these interactions can we explain many of the patterns and processes that occur in landscapes.

The landscape, the Earth's surface, is the stage on which ecology is played out. It comprises the landforms, the soils, the water and all the other physical features that influence the organisms that make up an ecosystem. And just as the landscape constrains and influences the ecology of a region, so too the ecosystems interact with and affect the landscape.

This book is about the profound but often subtle ways in which interactions affect both ecosystems and landscapes. Our aim is to help readers to understand the nature of complexity in the context of landscape ecology.

In the chapters that follow, we will explain what *complexity* is and what recent research has been learnt about it. We will also look at some of the many ways in which complexity turns up in ecosystems and in landscapes. As we shall see in Chapter 3, the landscape itself can be complex. In subsequent chapters we will look both at the many processes that make ecosystems complex, and at the ways in which the interplay between landscapes and ecosystems creates complexity of its own.

Finally, we will explore the relationship between landscape ecology, complexity and the information revolution. Besides describing some of the key ideas and the insights that flow from them, we will also introduce some of the techniques that are emerging to deal with ecological complexity in practice.

1.1   WHAT IS COMPLEXITY?

Like life itself, complexity is a phenomenon that is well known, but difficult to define. A general definition is difficult because the term complexity appears in different guises in different fields. In computer science, for instance, it usually refers to the time required to compute a solution to a problem. In mathematics it is usually associated with chaotic and other nonlinear dynamics.

Here we will take "complexity" to mean the richness and variety of form and behaviour that is often seen in large systems (Bossomaier and Green 1998, 2000). Complexity is not the same as size. For example, a herd of zebras feeding on a grassy plain do not behave in the same way as billiard balls on a table. If you strike one of the billiard balls it will roll around hitting the other balls. Eventually its motion and energy will dissipate through repeated collisions between the balls. This is simple behaviour. Although the balls interact, their energy soon averages out and they stop moving. On the other hand, if one zebra starts running, then the entire herd is likely to panic, creating a stampede. What is more, the stampede is not random. The running zebras avoid colliding with each other, but remain packed close together and head in the same direction. This is complex behaviour. The stampede emerges out of interactions between the zebras.

The property that is most closely associated with complexity is *emergence*. This idea is captured by the popular saying: the whole is greater than the sum of its parts. Emergence takes many forms. A forest emerges from the interactions of millions of individual plants, animals and microbes with each other and with the landscape. A forest fire emerges from the spread of ignitions from one plant to another. A flock of birds emerges from the individual behaviour of many individual birds interacting with one another. The organisation of an ant colony emerges from the joint behaviour of many individual ants interacting with each and with the colony environment. To understand complexity in ecosystems, we need to learn how large-scale properties like these emerge from interactions between individuals.

Which came first, the chicken or the egg? This famous conundrum exposes a gap in our intuition[2]. It is natural to assume that each cause has a simple effect, and vice versa. So an egg "causes" the chicken that hatches from it. Conversely, the chicken "causes" the egg that it lays. The chicken or egg question invites an answer in terms of simple causality. In reality, however, both the egg and the chicken are manifestations of a complex process[3]. Ilya Prigogine expressed this mental transition from static causal models to dynamic systems models concisely in the title of his book, *From Being to Becoming* (Prigogine 1980). Similarly, Barry Richmond talks about the need for structured thinking over simple causal thinking (Richmond 1993).

Many situations in landscape ecology are like the chicken and egg problem. Look again at the story that began this chapter. To understand what happened in the Amazon rainforests, it is necessary to realise that a rainforest is not a fixed object, but an on-going process. Traditional cause and effect classifications characterize a rainforest as a forest that grows in areas of high rainfall and soil that is rich in nutrients. Based on this model, it seems reasonable to conclude that if you cut down a rainforest for timber, then the high rainfall and rich soil will cause rainforest to grow back again in a few years. Not so! The truth is that the rainforest is a complex system. The species richness, the lush soils, and the high rainfall are all mutually dependent. They are each the product of a long feedback process. Higher-order or systems thinking is needed to understand the rainforest, which contains a vast network of feedback loops, flows and accumulations.

The tendency to think in terms of simple cause and effect leads to many problems in conservation. People see a local problem and seek a simple, local solution. They are often unaware of the spatial interactions that may be involved and do not realise what effects their local actions may have elsewhere.

The Murray River is Australia's largest river system. Its catchment, which and encompasses some of the country's most productive agricultural land, covers 14% of the continent receiving water from 41 tributaries across four states. The health of this river is of critical importance to the country's economy. However, in 2002, the National Trust of Australia was forced to declare the entire Murray River to be the country's most endangered heritage site: "Today no water flows into the sea from the Murray River. This once magnificent river now regularly fails to reach the sea." In 2002 the National Trust of Australia gave the following statement:

> *"The significant threats to the health of the Murray remain largely unaltered. However, the community engagement process, along with the work of community organisations have raised the profile of the Murray issue considerably. National action is required, and the National Trust*

---

[2] Philosophers have long been aware of problems with simple causality. For example, they distinguished between "proximate" and "final" causes. What causes a wild fire? A lightning strike may be the proximate cause, but hot weather and lack of rain also contribute as more final causes. These climatic conditions may themselves be caused by global warming caused by carbon dioxide emissions caused by oil and coal burning caused by humans. And so on.

[3] Evolutionary biologists would say that the egg came first. Fishes, amphibians and reptiles all laid eggs many millions of years before birds, and chickens, evolved.

urges the Council of Australian Governments to commit to a national approach to ensure that healthy flows are restored to the Murray, and that the community is fully informed and engaged in the conservation of our most precious resource – water." (NTA 2002).

Many factors motivated the Heritage Trust's decision. These issues included diversion of river water for irrigation causing floodplains and wetlands to dry up, dams causing permanent inundation of other floodplains, increasing pollution and sedimentation within the river itself, declining rainfall, salinization, desertification, and the spread of introduced species such as European carp throughout the system. Although data is limited, it is clear that the ecological consequences of these stresses are already severe. Populations of native fish, invertebrates and reptiles are in decline. Many species of waterbird have become locally extinct (Kingsford 2000).

No single person or organisation is responsible for the dying Murray. It is a *complex* problem that is emerging as the accumulated result of local actions and interactions all along the river. Essentially, the crisis has arisen because people tend to think of problems in terms of simple cause and effect. Farmers, councils, and government agencies have all acted to solve local environmental challenges without understanding the wider impacts that their actions then have on the river system as a whole.

1.2   WHAT MAKES ECOSYSTEMS COMPLEX?

When talking about complexity in the natural world, the discussion almost inevitably begins with biodiversity. Scientists use the term "biodiversity" in several ways — genetic variability, population and ecosystem diversity — but even if we take it to mean the number of plant and animal species, the global picture is enormous. Taxonomists have described about 1.7 million species, with 13,000 new descriptions per year (Kingsford 2000). The total number of species is estimated to be somewhere between 10 million and 30 million (May 1988). Most of the vertebrates and flowering plants are described, while invertebrates, bacteria and other microscopic creatures are largely undescribed (Wilson 1992; Tangley 1998). At the current pace of taxonomic research, it would take at least another 300 years merely to document the specimens already in collections (Green and Klomp 1997).

Although the number of species is impressive, it is not sheer numbers of species that make the living world complex, but the enormous variety of ways in which they combine and interact. Interactions between pairs of species can take many forms, such as predation, parasitism and competition. When there are many species present, the number of interacting pairs can be very large. For instance, suppose that 100 species inhabit a region; then there are 4,950 possible pairs of interacting species.

But species do not interact in pairs alone: the effect of one species on another may be altered by the presence of a third. When we look at possible combinations of multiple species, the possibilities blow out to astronomical proportions (Figure

1.1). There are over 1.7×10^{13} ways[4] in which communities of 10 species could form from a pool of 100 species. This number is about the same as the distance from the Earth to the sun in centimetres. For communities of 50 species at a time, this number rises to over 10^{120} combinations[5]. Within any given ecosystem, this complexity can increase by several orders of magnitude if we consider the possible interactions between organisms (biotic components) and their environment (abiotic components). These interactions determine the way an ecosystem functions. Every species is unique, and its behaviour forms a unique interaction with its environment. .

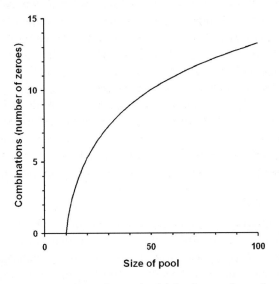

Figure 1-1. The increase (orders of magnitude) in the number of ways of drawing combinations of 10 species as the size of the pool from which hey are drawn increases. The vertical axis shows the number of zeroes in the number. For instance, the label 5 indicates the number 100,000 and 10 indicates the number 10,000,000,000 (ten billion).

Astronomical as some of the above numbers may seem, the reality is even more extreme. The very concept of interactions occurring between species is itself a simplification. Each species consists of unique individuals, and ecological interactions occur not between species as a whole, but between individuals belonging to each species. These interactions are governed partly by the spatial arrangement of individuals, making the landscape itself a major source of complexity.

The earth's land surface totals over 100 million square kilometres, but organisms are not spread evenly (nor in the oceans) because resources are patchy in both space and time. However, understanding landscape processes can make conservation easier. The identification of 'hotspots', or areas of unusual ecological importance, allows a more concentrated effort to achieve greater conservation

---

[4] The number 1×10^{13} (ten trillion) is written as 1 followed by 13 zeroes.
[5] A number greater than the number of atoms in the universe!

results (Pimm and Raven 2000). At least 25 such hotspots are known to exist around the world (Myers et al. 2000). They contain the highest concentrations of endemic species[6] and are also experiencing significant loss of habitat. Packed within just 1.4% of the land surface area on earth, these hotspots contain 44% of all threatened vascular plant species and 35% of the threatened species of amphibians, birds, reptiles and mammals.

Given the richness of ecological interactions, we cannot hope for a simple and general understanding of the behaviour of ecosystems, such as we have of (say) the behaviour of gases. However, there is another way to approach the problem. This approach, which sees ecological interactions as a network of connections, recognises that although complex systems are diverse, they also tend to share certain internal structures and processes that lead to consistent behaviours.

At the simplest level, connections form chains. Foxes eat rabbits; rabbits eat grass. These connections constitute a chain which creates an indirect effect of foxes on grass. More generally, suppose that species A eats species B, that species B eats species C, and that species C eats species D. These species form what is known as a *food chain*, where each species exerts an influence on other species further down the chain. Often these chains of interaction are circular. The number of foxes influences the number of rabbits, but the number of rabbits also influences the number of foxes. These circular interactions, known as feedback loops, play a key role in ecosystems.

Negative feedback occurs when activation of one part of a system reduces the activity of the component that activated it. Negative feedback loops result in convergent behaviour. For example, increasing numbers of rabbits might lead to increasing numbers of foxes, which in turn would lead to reducing the numbers of rabbits again. In theory, negative feedback would stabilise ecological systems. The fox and rabbit populations converge on numbers that are in balance. Any small, random change in the number of rabbits would instantly be compensated by an equivalent change in the number of foxes (and vice versa), so the numbers of rabbits and foxes would remain roughly constant. In practice, the response of one component to another is often delayed. This can lead to oscillations in population size, and to a wide variety of non-linear and chaotic behaviour (May 1976). Chaotic changes in population size occur when the time for response of reproduction by prey to changes in predation is much shorter than the response of predators to prey population.

Sometimes, activation of a system component increases the activity of the component that activated it. We call this positive feedback. It results in divergent behaviour. Rolling snowballs downhill, compound interest, and nuclear explosions are all examples of positive feedback.

In ecology, we see positive feedback loops wherever species compete with one another. An increase in one species leads to a decrease in its competitor, leading to a further increase in the first species. Positive feedback is often destabilizing, causing chaos and the destruction of complex systems; paradoxically, as we shall see later, it can also be an important means of creating order.

---

[6] Species that are found nowhere else.

As emphasized above, the notion of predator-prey relations or even food chains is a vast oversimplification. The reality is that there are multiple prey and predators, herbivores, producers and decomposers that form complex food webs (see Chapter 7), which may exhibit high levels of redundancy or high levels of compensatory feedback behaviour.

We mentioned earlier that the landscape itself is a source of complexity in ecosystems. Spatial features, such as soil types, temperature and humidity gradients, topography, wind and water currents, connect sites in a landscape in many different ways. These connections form diverse patterns which influence the distributions of plants and animals. Spatial processes, such as seed dispersal and migration, provide mechanisms by which different sites in a landscape interact ecologically[7]. In later chapters we will explore the major factors that govern these spatial processes.

If patterns in space are important, then so too are patterns in time. Apparently random events, such as drought, flood and fire, each cause massive and unpredictable mortality, but also provide opportunities. Both the frequency and magnitude of environmental variations can strongly influence ecological structures. It is important to remember that the ecosystems can affect landscape characteristics, such as soil stability or rainfall and temperature. Preserving Amazonian forests prevents soil erosion, whereas large scale deforestation can change rainfall and temperature patterns.

Understanding the interaction between feedback processes and temporal and spatial variation within ecosystems has helped to resolve some key questions that cannot easily be approached empirically. For example, ecologists noticed that ecosystems with very large numbers of species, such as the Amazon rainforest, tended to be very stable, with low extinction rates and little variation in population sizes over time. In contrast, they saw ecosystems with only a few species, such as Arctic tundra, experiencing much larger fluctuations in population sizes and frequent extinctions. This led ecologists to believe that complexity in ecosystems enhances their stability, making them more robust to disturbances.

However, during the 1970s a series of models showed that exactly the opposite was likely to be true[8]. Simulation models in which sets of interacting populations are formed randomly predicted that systems with greater numbers of species are more likely to collapse, simply because there is a greater chance of forming positive feedback loops[9]. This finding suggests that the arrow of causality points in the opposite direction. Rather than complexity creating stability, stability permits complexity. It is no coincidence, then, that stable and complex ecosystems occur in areas of minimal seasonal variation near the equator, while instability and simplicity are associated with the extreme seasonality of the Polar Regions.

Similarly, one of the most important results to spring from ecological studies is that environmental factors alone do not fully explain the distributions of organisms. For instance, competition between species often truncates distributions

---

[7] We will look in detail at the effects of spatial patterns and processes in chapters 3, 6, 7 and 8.
[8] The most influential of these studies was work by Robert May (1973).
[9] See, for example, Tregonning and Roberts (1979). We look at feedback again in Chapters 4 and 6.

along an environmental gradient (Pielou 1974). An epidemic may reduce the size of a population to a point where inbreeding decreases survival and reproduction (Schaffer 1987). All of these results imply that ecosystems are not controlled in a simple fashion by either external or internal factors, but by the interaction of many biotic and abiotic factors (e.g. Bull and Possingham 1995).

1.3   WHY STUDY ECOLOGICAL COMPLEXITY?

In September 1993, a group of eight men and women emerged into the open air, after living for two years in a closed, artificial habitat. Their home for all that time had been an interconnected series of domed greenhouses located in Arizona's Sonoran desert, about 30 miles north of Tucson. Called Biosphere 2, the $150 million experiment had two main goals. Firstly, NASA, the USA's National Aeronautics and Space Administration, wanted to test the viability of colonies in space. Secondly, isolated ecosystems provided a chance to learn more about issues for the long-term maintenance of Biosphere 1, the Earth itself. The two year experiment provided many valuable lessons. Chief among these lessons was the realization that managing a closed ecosystem was a far more complex and delicate problem than anyone had imagined[10]:

> *"Artificial biospheres of the scale and complexity of Biosphere 2 can only work with coordinated rigorous design at each level of ecology: biospheres, biomes, bioregions, ecosystems, communities, patches, phases, physical-chemical functions, guilds, populations, organisms, and cells (both eukaryotic and prokaryotic).*
>
> *The coral reef proved remarkably responsive to changes in atmospheric composition, light levels and climatic conditions, requiring skilled and frequent management intervention."*

Certain differences from the Earth's biosphere caused problems in the habitat. For instance, the ratio of carbon held in living biomass to carbon held in the atmosphere is about 1:1 on Earth as a whole, but in Biosphere 2 it was about 100:1. Likewise, the residence time for carbon dioxide in the atmosphere was about 3-4 days in Biosphere 2, hundreds of times less than for the Earth's biosphere, which has a vast buffer of air. This meant that the carbon balance was highly sensitive to changes in any part of the ecology.

More subtle issues arose from the self-organisation that occurred within the system:

> *"Examination of the ecosystem within Biosphere 2, after 26 months... [showed] features of self-organization....the self-organizing system*

---

[10] Cited from Allen (2000) and Allen et al. (2003) respectively. Other accounts include Cohen and Tilman (1996).

appeared to be reinforcing the species that collect more energy ..."
(Odum 1996)

Unexpected difficulties emerged, and the inhabitants had to intervene to prevent runaway problems from taking hold. For instance, oxygen levels dropped dangerously low, requiring repeated input from the outside world (Severinghaus et al. 1994). Researchers suspected that rich soils chosen to promote plant growth were causing uncontrolled growth of soil microbes, whose respiration consumes oxygen. But that created a further puzzle: if soil microbes were responsible, why did atmospheric carbon dioxide levels remain constant? The missing factor was eventually identified as unsealed structural concrete absorbing carbon dioxide. If a stable environment proved unattainable even in the small, simple and carefully planned world of Biosphere 2, then how much greater is the problem of managing the Earth's ecosystems?

We need to study ecological complexity. The solution of many crucial environmental problems hinges on our understanding of complexity. With modern global trade systems, the human world is now so interconnected that potentially everything can affect everything else. The echoes of human activity are felt everywhere, even in regions usually regarded as wilderness. Furthermore, we cannot deal with ecosystems in isolation. To manage ecosystems and species in one area, we need to understand their place in the larger environmental context on a global scale. Conservation in many countries is really a question of economics, and chains of events link processes that cross national boundaries. Deforestation in the Amazon is driven by demand for cheap hamburgers in the United States. Political instability in the Middle East raises fuel prices worldwide, causing greenhouse gas emissions to drop briefly. War in the Middle East causes massive increases in greenhouse emissions when oil wells are set on fire. A hydroelectric dam that will power a whole province in China will drive an endangered dolphin to extinction. Saving a rare orchid in Queensland may lead to a US mining company losing millions of dollars.

The problem is urgent. Environmental management has become a race between conservation and exploitation. On the one hand, we need to develop an understanding of ecosystems and to set in place the necessary agreements, infrastructures and practices. On the other hand, environmental exploitation and degradation (especially land clearance) are now reaching into environments everywhere.

To understand ecology, we need to understand complexity. In the next chapter, we shall look at the sources of complexity in ecology and at some of the ways in which ecologists have sought to understand it.

1.4   A NEW ECOLOGY FOR A NEW MILLENNIUM?

Many people still alive can remember the early days of computing. In the 1950s, the world's most powerful computer was a huge machine. Built using thousands of tubes and valves, it filled a large room. Today I can place in my pocket a computer that is about a million times faster, that holds a million times as much data, and is a million times smaller than those early machines. Along with the increase in raw power, the range and diversity of computers has blossomed. Not only does virtually every home and business in the western world own a computer, but also we have computers in our cars, watches, phones, TVs and dozens of other everyday devices.

The above figures highlight the magnitude of the information revolution that has transformed society since the mid-twentieth century. The revolution is also transforming science. In the 1950s, data as a commodity was rare and expensive; whereas today it is abundant and cheap. This transition marks a huge change in the way science is done. Computers today provide landscape ecology with a host of tools, such as geographic information systems (GIS) to work with landscapes, databases and data warehouses to store environmental information, and virtual reality to simulate environments. Advances in communication mean that these resources are now available online, so that everyone, everywhere potentially has access to 'e-science'.

The information revolution is also changing the way we perceive the world around us. Every age projects its preoccupations onto the world around it. The Inuit people, for example, have some 16 words for snow[11]. In the course of the Industrial Revolution, especially in the nineteenth century, machines and mechanistic views of the world dominated people's thinking. As a result, ideas based on machines dominated science. The world was seen as a great machine. The planets moved around the sun like clockwork. Living things were seen as factories, with a power source and systems for transport, waste disposal, and central control.

In the information revolution, a new view of the world has emerged. There is a huge gulf between the perceptions of the "Net Generation" and all previous generations (Oblinger and Oblinger 2005). Nature today is often seen as a great computer whose programming we have to discover. This idea of natural computation fits well with biology. People often compare the genetic code to a computer program because it contains the "recipe" for building new organisms. Ribosomes, the cellular machines that make proteins, are information processors in that they "read" genetic information transcribed from DNA and output proteins according to its instructions. Sensory perception involves processing data input. Animal communication involves the transfer of information from one animal to another, and the subsequent interpretation (processing) of that information. Animals often act as though they are programmed to behave in certain ways.

Not only can nature be regarded as computation, it is no exaggeration to say that advanced computing is becoming more and more indistinguishable from biology. This is happening because computing deals increasingly with problems that resemble biological or ecological systems. For example, in the drive to increase processing power, computers have gone from single chips to multi-processors, then to distributed processing, and eventually to swarms of processing agents.

---

[11] Source: *Inuit Made Easy* http://www.geocities.com/~westracine01/Nation-InuitLanguage.htm

To solve the increasingly complex challenges posed by such problems, computing has looked to living systems, which have evolved ways of coping with complexity. The result has been a proliferation of biologically inspired ideas borrowed by computing. These include such ideas as cellular automata, genetic algorithms, neural networks, and swarm intelligence. We shall see more about these ideas in later chapters.

While it is risky to take computing analogies too literally, the idea of natural computing does help to explain certain aspects of ecology, as well as many other natural phenomena. What is stimulating these new ways of viewing nature is the recognition that new methods are needed if we are to understand the complexity of the living world. By themselves, traditional ways of doing science are not enough to allow us to fully understand complex systems and processes. Trying to come to grips with complexity poses a challenge for science. The reason for this is that complexity is about the way the world is put together. However, the traditional way of doing science is to do the exact opposite: that is, to take things apart and examine each piece in isolation. This is known as the reductionist approach.

The reductionist approach has served science well. The wonders of modern technology are tangible proof of the success of the method. The principle is simple: "divide and rule". If you want to understand something complex, then break it down into its component parts and figure out how they work. If necessary, keep dividing until you reach a level that is simple enough to understand.

This is an immensely powerful idea. Over the course of four centuries, physicists and chemists have dissected matter down into its components: molecules, atoms, protons, electrons and, more recently, bosons and fermions. At every step of the way, they made important discoveries that have helped us to understand aspects of nature as diverse as the origins of the universe and the structure of DNA. These discoveries led to all manner of practical applications, from aeronautics to biotechnology.

There is a well-established logic behind the reductionist approach. For instance, if you want to know how a plant performs under different conditions, then put it in growth chambers and take precise measurements of respiration, photosynthesis and growth. What you find are clear patterns of cause and effect. If you raise the temperature by so many degrees, then you can observe the resulting effects on photosynthesis. If you alter day length by a certain amount, then the results are not only repeatable, but also predictable. With this information, not only can you understand why the plant grows better in certain environments, but also you can predict exactly how it would fare in any other environment.

Unfortunately, the world does not always divide itself in straightforward ways. When you pull things to bits, you learn about the parts, but unexpected things can happen when you put them back together again. For instance, by experimenting with individual plants in a growth chamber, you may learn how that plant responds to light, water and so on. But when it grows out in the wild, it interacts with many influences, especially other plants. Laboratory results are often not repeatable under field conditions. No laboratory experiment can account for every factor that might affect plants and animals in the field. Interactions do matter. Often, the whole is much greater than the sum of its parts.

Learning what happens when you put things back together is what complexity research is all about. Such questions abound in science. How do billions of neurons become organised into a living brain? How does the genetic code control the growth of developing embryos into complete human beings? How do ants and bees manage to construct elaborate nests and societies without the ability to think and plan?

In many fields, scientists have found that they needed to develop new ways of dealing with the issues that arise when many things interact with each other. Physicists, for instance, needed to understand what happens when large numbers of particles, such as atoms, or stars, interact. The need to understand the whole system does not mean abandoning the tried and true divide-and-rule approach of reductionism. It means that traditional approaches need to be complemented with new methods and ideas. In brief, as well as understanding the parts, we also have to understand how the whole emerges from those parts.

Ecology has developed many ideas, such as diversity and food webs[12], that recognise the inherent complexity of ecosystems. What the information science revolution has done is to provide new tools that can deal with complexity in more direct ways than were previously possible.

By allowing us to deal with complexity more directly, the information revolution provides new ways of doing ecology, which complement traditional approaches. Features of this new science of complexity include the following:
- it is based on the idea of natural computation, interpreting natural processes as computation;
- computers play a crucial role, both in their ability to handle large volumes of data, and in their ability to create "virtual worlds";
- for complex systems, no single model can be all-embracing, so different models may be needed for different purposes, including explanation, prediction, and control;
- complex systems often behave unpredictably, so we need to study scenarios (that is, to ask 'what if' questions).

In the chapters that follow we will explain these ideas in more detail. Moreover, we will provide an overview of what they can tell us about landscape ecology. We will explain many of the methods used and summarize some of the fresh insights that they are providing about ecosystems and environmental management.

Only in the past few decades has science begun to make inroads into understanding complexity. In the early part of the twentieth century, there was a mood among scientists that complete knowledge of how the world works was just around the corner. Physicists had penetrated matter to its core, discovering the atomic building blocks of all matter. At the beginning of the third millennium, there is no such mood of optimism. Instead, there is acute awareness of just how little we yet know. The greatest challenge is to understand the complexity of the living world.

---

[12] We discuss diversity in chapters 4 and 10 and food webs in Chapter 7.

CHAPTER 2

SEEING THE WOOD FOR THE TREES

A rainforest emerges from complex interactions between the many different plants and animals that comprise it.

The whole is greater than the sum of its parts. As we saw in chapter 1, this popular saying captures the essence of complexity. In this chapter, we look briefly at examples of processes in which fine scale functions, at the level of individuals, lead to larger scale organisation. Many other issues (e.g. foraging strategies) arise from individual interactions besides the examples given here. We will meet some of these in later chapters.

Sometimes you are better off not to know too much. If you are trying to understand the workings of a flock of birds, or of a forest ecosystem, then you can only get so far by studying the individuals. Of course, you do need to know something about the individuals that go to make up the whole. Some details matter. The way that living creatures interact, the properties that allow one plant to outgrow another, the relative speed of predator and prey are all characteristics that count on a larger scale. But there are diminishing returns in probing individuals ever more deeply. Once you understand the essential workings of metabolism, for instance, further detail will probably not help you to understand how that animal will fare in its struggle for survival.

In trying to understand how self-organisation works, how a large number of individuals become organised into an ecosystem, a certain amount of abstraction is needed. Traditional models tend to gloss over processes that involve interactions at the level of local individuals or elements. Instead they tend to look either at the large scale, or else at the fine details. They tend to take a top-down view of how constraints act on individuals, rather than a bottom-up view of the effects that arise from interactions between individuals.

In so doing, they get only half the story. They deal with how plants grow and how animals move. In both cases, the individual-oriented models take bottom-up views that offer us very different insights from top-down, reductionist models. This is not to say that we should abandon the traditional models and use only individual models. Our point is that to gain a complete picture, we really need to look at both.

2.1   PLANT GROWTH AND FORM

The growth and form of plants helps to determine the ways in which they interact with their environment, and with other plants. Leaves are angled to catch the sun (or to avoid it). Tree trunks and branches are often structured in such a way that water is funneled to the roots. The shapes of plants play an important part in determining their survival in the landscape and their role within an ecosystem. The main impact of interactions between different plants, and between plants and their environment, lies in the ways they affect growth. The supply of sunlight, water and nutrients all affect the rate of growth. In turn, the nature of a plant's growth helps to determine the way it interacts with its environment, and with other plants. So to understand where plants are found in landscapes, and in what numbers, it is essential to understand something about how they grow. Shade-tolerant plants grow slowly, but they can survive under the canopy of other trees. They often cluster leaves around the outside of the canopy to maximise the amount light they collect. Fast-growing seedlings quickly outgrow their competitors, but they need lots of light and prefer open conditions.

There is potentially much more still to learn about the nature of plant growth and the interplay between plant form and function within landscapes. There is strong motivation for looking more closely at how growth occurs. At the time of writing, biotechnology has already provided complete maps of the genetic makeup of several species. This next step is to unravel the mechanisms by which those genes control growth. In other words, the challenge for research is to cross the traditional boundaries imposed by scale, and to understand how genes affect growth and how growth affects geographic distribution.

Traditional models of plant growth have usually focussed on large-scale aspects of growth (Blackith and Reyment 1971). Descriptions of plant form can produce elegant, mathematical descriptions of plants as snapshots. However, they tend to divorce form from function. The manner of drawing the plant does not reflect the processes involved in morphogenesis. These descriptions embody no information about the change in size and form with time.

Causal models have usually dealt with simple attributes of growth, such as total biomass or canopy area, and tried to relate them to other variables. However, reduction of a landscape's vegetation to unit area or volume relies on simplifying assumptions about the distribution of plant parts in order for analytical solutions to be tractable. For example, by assuming an even spatial and angular distribution of leaf segments, it is possible to calculate quantities such as light interception[1]. Again, this approach relies on a snapshot. This snapshot can be used within a growth algorithm, but no account is taken of the leaf distribution on the pattern of growth.

Amongst the current paradigms for modelling growth, perhaps the most prevalent approach, both in plants and animals, is what we might term reaction-diffusion systems. Models of this kind look at growth as a process that consists of biochemical reactions. Most importantly, the rate at which these reactions occur, and therefore growth rate, is governed by the diffusion of substances across biochemical gradients. In essence, these models consider the way external controls and constraints affect growth rates. The rate at which a plant grows, for instance, is limited by the rate at which water and nutrients can be supplied to the growing tissues. This is very useful to know; however, it is far from the complete picture. What is missing is the detail of how the growth is organised. It is this problem that we turn to now.

2.1.1 Branches and leaves

The most important features about the organisation of plant growth are that it is *modular*, and it is *repetitive*.

By *modular* we mean that the growth consists of parts that we can treat as distinct units. The most common of these units are branches and leaves. The modularity is important. Within a given leaf, for instance, we do not need to worry

---

[1] Monsi and Saeki (1953) used a negative exponential formula.

about all the internal details. We can forget about all the individual cells. In understanding the growth of the entire tree, all we need to know is that the leaf has certain attributes (e.g. size) and that it operates in a particular way.

By *repetitive* we mean that the same processes repeat themselves over and over. When a branch forms, it grows and other branches and leaves develop from it. In their turn, those branches grow and other branches form, and so on.

Together, these two properties – modularity and repetition – make patterns of plant growth self-similar. A fern frond, for instance, is made up of segments that are smaller versions of the whole. If you look closer still at those segments, each, in turn, appears to be made up of smaller versions of the larger. The pattern repeats itself. In this sense, plant growth patterns form natural *fractals*[2].

We can capture repetitive growth patterns, such as branching, as rules. Let's start with a simple example. Suppose that we have the following two rules that tell us how to replace one symbol with another:

$$A \rightarrow B$$
$$B \rightarrow AB$$

What Rule 1 means is that if you have the symbol A, then you replace it with a B. Likewise Rule 2 means that if you find the symbol B, then you replace it with the pair of symbols AB. Now suppose that we start with the symbol A. Then by repeatedly applying the above two rules, we can generate strings of symbols of ever increasing length. The following list summarizes the sequence of results at various stages, starting from stage 0, where we have just the initial symbol A.

```
Stage 0 : A
Stage 1 : B
Stage 2 : AB
Stage 3 : BAB
Stage 4 : ABBAB
Stage 5 : BABABBAB
Stage 6 : ABBABBABABBAB
Stage 7 : BABABBABABBABBABABBAB
```

An interesting feature of this sequence is the lengths of the strings. If we count the symbols at each stage then we obtain the sequence:

$$1, \; 1, \; 2, \; 3, \; 5, \; 8, \; 13, \; 21, \; 34, \; \ldots.$$

Some readers may recognise this as the famous Fibonacci sequence of numbers, which has well-known associations with plant growth. What this model shows is that the sequence is a natural consequence of the iteration process.

The above model is an example of a formal language model, in which the rules for rewriting string define the grammar (syntax) of the language. Formal languages

---

[2] See chapter 3.

are now widely used to model the organisation of growth processes. Such models are usually called *L-systems*. They are named after Aristid Lindenmayer, who was one of the first people to use syntactic methods to model growth (Lindenmayer 1968). The power of L-systems comes into play when we assign meaning to the symbols and rules.[3]. For instance, if the symbols represent branches of a growing tree, the rules would denote the organisation of branching.

The central idea of L-systems is that of rewriting, a technique for constructing complex objects by successively replacing parts of simpler, initial objects using rewriting rules. L-systems are distinct in their application of rewriting rules in a parallel manner. All predecessor symbols are replaced with successor strings simultaneously. This procedure allows us to capture simultaneous processes of expansion, division and initiation of several different cells or plant parts at the same time.

$$
\begin{array}{rcl}
X & \rightarrow & F[A]FY \\
Y & \rightarrow & F[B]FX \\
A & \rightarrow & X \\
B & \rightarrow & Y
\end{array}
$$

The brackets here refer to modules that represent branches. That is, the string within any pair of brackets represents a pattern of growth within a branch off the main stem. If we apply the growth rules several times, then it leads to the following set of strings:

Stage	Pattern
0	X
1	F[A]FY
2	F[X]FF[B]FX
3	F[F[A]FY]FF[Y]FF[A]FY
4	F[F[X]FF[B]FX]FF[F[B]FX]FF[X]FF[B]FX
5	F[F[F[A]FY]FF[Y]FF[A]FY]FF[F[Y]FF[A]FY]FF[F[A]FY]FF[Y]FF[A]FY

Notice how each symbol gets replaced by the symbol, or string, that is given in the rules. If we were to continue the process, then as in growth, the string just keeps on getting bigger and bigger. It is very difficult to make any sense of strings of symbols like these. However, we can translate the strings into a picture that shows what the resulting structures look like. If we represent each symbol F as a line segment[4] here, then we can translate each symbol into a picture element. We need to make an important distinction between the symbols A and B. We take A as denoting a branch to the left of the main stem, and B to denote a branch heading off to the right of the stem.

The resulting growth pattern is shown in Figure 2-1.

---

[3] The process often involves using several different rules in parallel, not just one at a time.

[4] This drawing process uses turtle graphics, which we look at later in the chapter.

Figure 2-1. The branching pattern produced by the L-system given in the text. Notice that the pattern is self-similar; that is, each branch and sub-branch is similar in form to the entire pattern.

Here we have a minor confession to make. The picture in Figure 2-1 is not quite the pattern generated by the model. If you look closely, you will see that the line segments near the start of each branch are longer than the segments near the tip. This is a slight refinement of the above model, in which we incorporate an expansion factor to allow for continued growth of stem segments after they first appear. Another refinement is to incorporate a brief delay before the expansion of each branch occurs. We have omitted such details from the above description to keep the model simple.

The important thing about this model is that it highlights the modular and iterative nature of plant growth. It also shows that we can capture the apparent complexity of plant growth in a surprisingly small set of simple rules.

2.1.2 Overall plant form

We have seen that formal languages can capture the organisation of many growth processes very simply. Therefore, models based on formal language can help us to understand certain implications of plant growth and form. Consider the growth patterns of the two (hypothetical) trees shown in Figure 2-2 and governed by the following simple L-systems:

    (1)   A → [0]A[0]
    (2)   A → [A]A[A]

Here the symbol A denotes a growing tip, and 0 denotes either a terminal or a slow-growing side branch. As before, square brackets indicate branches. In the model, the symbol 0 remains constant. The first few growth stages of these models are given by the following sequences:

Stage	Model 1	Model 2
0	A	A
1	[0]A[0]	[A]A[A]
2	[0][0]A[0][0]	[[A]A[A]][A]A[A][[A]A[A]]
3	[0][0][0]A[0][0][0]	[[[A]A[A]][A]A[A][[A]A[A]]] [[A]A[A]][A]A[A][[A]A[A]] [[[A]A[A]][A]A[A][[A]A[A]]]

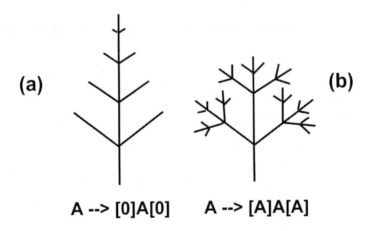

(a)   (b)

A --> [0]A[0]   A --> [A]A[A]

Figure 2-2. Plant growth patterns arising from the rules shown. In the same length of time, the plant (a) grows taller, but has fewer branches and leaves than plant (b).

The point of this model is to demonstrate how these two different patterns affect growth. What happens in practice is that the trunk supplies water and nutrients to the branches. (Let's ignore for the moment that the roots grow too, and assume that both trees have identical root systems and both trunks deliver nutrients at roughly the same rate). Now if we make the reasonable assumption that a certain volume of nutrients is needed to produce each branch element, then the supply of nutrients up the stem limits the rate at which development proceeds. In other words, the two trees will progress through the developmental stages shown above at different rates.

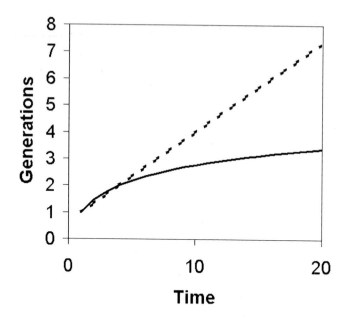

Figure 2-3. The relationship between branch "generations" and time in the two models shown in the previous figure. The dashed line corresponds to plant (a) and the solid line to plant (b).

   The two development curves are illustrated in Figure 2-3. As the development curves show, the first tree produces higher orders of branches at a constant rate, but the growth rate of the second tree continually declines. This means that, when the two trees reach maturity, we can expect the first one to have a tall crown, whereas the second will a lower crown with a larger diameter, forming a dense bush[5]. These different shapes in turn will have implications for the survival and reproduction of the trees in different environments. The bush may do better, for example, in a windswept desert where its shape offers little resistance to the wind, while the tall tree might thrive as a fast-growing colonist monopolising an opening in a forest canopy.
   Admittedly, the above models are extremely simplistic. For example, the rate of supply to a tree of water and nutrients increases as its root system grows and spreads. Also, not all nutrients and energy go into growth—metabolism and reproduction require many resources too. However, despite being so crude, the models do highlight the consequences of two extreme strategies that trees can adopt in evolving their growth patterns. The point is that a simple change in the

---

[5] If the numbers of branches on the two trees at time t are $B_1(t)$ and $B_2(t)$, Then according to our assumptions,
$$B_1(t) = m/3\ (2kt\text{-}1) \text{ and}$$
$$B_2(t) = m\ (2kt\text{+}1),$$
with m and k being growth constants.

pattern of growth leads to a large difference in the ways that the two plants will fare in a landscape and in competition with other trees. Even individuals of the same species will change their development patterns and the resulting shape depending on planting density. Dense plantings will result in taller trees with less side branching. If the same tree is planted in isolation, it will have many more side branches and a more spheroidal canopy.

Growth models of this kind raise the prospect that it may be possible to link patterns of development to genetic traits, and by looking at the effects of those genetic traits on growth, to relate them to the environmental conditions that are favourable to the plants.

2.1.3 Self-organisation versus constrained growth

Grammatical models of complex growth processes, especially those found in plants and plant tissues, have highlighted the importance of communication between cells[6]. One criticism that can be made of grammatical growth models is that *ad hoc* rules often need to be included in the syntax to make the system work. For instance, the algal growth model (Figure 2-4) requires that cells pass through intermediate states between divisions. In the model, these intermediate states ensure that branching occurs at the right stages, but their physiological and biochemical meaning is undefined. While it is possible that intermediate states are related to unknown control mechanisms, it is just as likely that they have no biological meaning at all.

Figure 2-4. Cell lineages in the alga *Chaetomorpha linum.*[7]

Botanists often express two concerns about L-system models. Are they valid? And are they useful? Validity was a genuine concern with some of the earliest models. Some of the early models, which were strongly influenced by linguistics and formal language theory, focussed on context-free models, whereas most growth processes, not to mention most biological processes, are clearly context-sensitive. Notice that in the models above, the rules only ever contain one symbol on the left

---

[6] This means that context-sensitive grammars are often required to describe the processes adequately.
[7] Based a model described by Herman and Rozenberg (1975).

they are complementary models. Reaction-diffusion models explain the effects of the biochemical environment or context, while the L-system models explain the fine-scale details of the process. A complete picture of plant growth (the same argument applies to animal growth as well) requires models that combine the two approaches.

2.2   ANIMAL MOVEMENT

From a practical point of view, animals are much harder to study than plants. Plants just sit there; animals move around.

At large spatial scales, the distributions of plants and of animals present similar problems. However, at the local scale, the issues involved are very different. Some of the great challenges of ecology are to identify processes that govern the ways animals behave and to understand the effects that arise from their interactions with each other and with their environment. As we have seen previously with other kinds of complexity, these local interactions often have global effects.

In this section, we shall look at some of the models that have been developed to address the complexity of animal behaviour, especially phenomena that emerge from interactions between many animals.

To begin with, we need a simple, abstract way of representing animals. Let's start with a well-known model: a robot. A robot is essentially a computer that can move around, sense its surroundings, and do things. When we say that it is a computer, this means that its behaviour is governed by a program of some kind.

2.2.1 Turtle geometry

Perhaps the simplest behavioural model based on a robot is *turtle geometry*. Admittedly this is a model of geometry, not of any real animal's behaviour. However, it does serve to demonstrate the key principle that we can capture many kinds of complex motion with simple rules of behaviour. It also shows how we can begin building more sophisticated models.

In the 1970s, the Stanford mathematician Seymour Papert devised a drawing system called turtle geometry (Papert 1973). His aim was not to understand turtles, but to help children to learn mathematical ideas by embodying them in concrete form. The idea of turtle geometry is to think of a geometric shape as the trail left by an imaginary turtle as it wanders around on a surface. If we were to attach a pen to a turtle and place it on a table covered with paper, then the pen would draw a line marking where the turtle has been.

The turtle's path can be described by a sequence of symbols representing its actions as it moves around. In the simplest case, there are just four possible actions: Step Forward, Step Back, Turn Left, and Turn Right (both turns through 90 degrees). Let's abbreviate these using the symbols F, B, L, and R, respectively. Forget, for the moment, that a turtle is a complex animal with four legs and so on. All that matters is that it *can* move forward and do all the rest. The details do not matter.

Given the above symbols, we can describe a trail in terms of the sequence of actions that the turtle makes as it moves, such as FFRFFFLFRFF. Strings of symbols like this one form a simple language. Not only can we use the strings to represent a path after the fact; we can also use them to program a robot turtle to follow a certain path, and, in the process, to draw a particular shape. For example the string FRFRFRFR tells the turtle to draw a square.

The above idea becomes powerful when we introduce names for particular actions. For example, suppose that we assign the name SQUARE to the string given above for drawing a square. Then instead of repeating the detailed list of actions each time, all we need do is to give the turtle the command "SQUARE" and it will automatically carry out the sequence FRFRFRFR. In this way, the turtle's repertoire of behaviour grows. What is more, we can build up ever more complex patterns out of the elements that it has already learned. So for instance, by adding the following two rules, we can teach it to draw an 8×8 grid of squares.

```
ROW   → 8 SQUARE L 8F R
GRID → 8 ROW
```

Of course, to produce really refined behaviour, the turtle needs to be a bit more skilful to begin with. Most versions of turtle geometry allow movements forward by fractions of a step, and turns by any number of degrees. Given these, and a few other enhancements, we can encapsulate a bewildering variety of patterns as turtle programs. Turtle geometry is often used to draw plant growth patterns of the kind that we saw earlier in this chapter.

2.2.2 From turtles to agents

The grid example above embodies an important idea in computing, and in behaviour: *modularity*. Modularity reduces complexity. Complicated behaviour is difficult to learn, just as complicated computer programs are difficult to write. If we were to sit down and try to write out a string to describe the turtle's complete path in drawing the honeycomb pattern, then it would be a long and difficult process. The chances of making a mistake are great. By giving names to elements of the desired pattern, we carve a big, complex problem up into simpler modules. Once we know that a particular module does its job correctly, then it becomes a building block that we can use again and again.

Let's bring the idea of modularity a bit closer to home. What happens when you get up and go to work in the morning? Most people would produce a list something like this:

```
get out of bed
have a shower
get dressed
have breakfast
```

```
pack my briefcase
leave home
catch the bus
```

The details will vary from person to person, and even from day to day. But we can all identify with this sequence of events. But look more closely. Each event is really a routine, a behavioural module, that encapsulates an entire sequence of activities. Now let's look at one activity, such as having breakfast. We can likewise carve up this episode into a routine sequence of actions. It might look like this:

```
get out cereal, milk and cutlery
fill bowl with cereal and milk
heat bowl in microwave
sit and eat cereal
clear breakfast table
```

The modularity does not stop there. We could go on. "Heating the bowl in the microwave", for instance itself involves a sequence of steps, such as this:

```
open microwave door
place bowl inside
close door
set timer
press start button
remove bowl from microwave
```

and so on. And each of these actions could also be broken down into a sequence of hand-eye movements.

The point of this example is that our own behaviour is highly programmed. We are so accustomed to thinking in terms of behavioural modules that mostly we do not even realise that we are doing it. In the same way, we can summarize the behaviour of animals in modular terms.

To move from geometry to models of real animals, we need a slightly more sophisticated model than Papert's turtle. The model's components are usually called agents. An *agent* is an object that can interact with its environment. The simplest agents are like the robots above. Other common attributes often (but not always) assigned to agents are:

- computational intelligence (i.e. knowledge expressed as rules),
- goal-directed behaviour, and
- the ability to interact with other agents.

A common finding in agent models is that the behaviour of a large-scale system emerges from the properties and interactions of many individual agents.

2.2.3 The boids and the bees

In the early 1980s, two Dutch investigators, Paulien Hogeweg and Brian Hesper, began to look at what happens in systems in which many simple agents interact. One of their earliest successes was a simulation model of bumblebee colonies (Hogeweg and Hesper 1983). In this model, the bees are represented as agents that behave according to a set of simple rules. There was no set plan of action for each bee. They simply obeyed what Hogeweg and Hesper called the TODO principle. That is, they moved around, and in doing so, they simply did whatever there was to do in their scheme of behaviour. What Hogeweg and Hesper found was that the social structure that is observed in real colonies emerged in the model as a result of interactions among the virtual bees with each other and with their environment.

The above result is important. It shows that order can emerge in a system without any planning or design at all. Bees are not intelligent. They have no concept of what the overall structure of a colony should be. Instead, the order is a by-product of the complexity in the system. It emerges out of a multitude of interactions.

Subsequent research has extended the bumblebee result to many other kinds of animals, from simulated ants to virtual birds ("boids"). Ants, for instance create nests by wandering around and obeying simple rules, such as, "if you find a stray egg pick it up; if you find a heap of eggs then dump the egg you are carrying" (Figure 2-6). By obeying this simple rule, the ants quickly sort the eggs into piles (see Figure 2-7). We call this behaviour an "ant sort". Ants achieve other feats of organisation, such as foraging for food, by following other sets of simple rules.

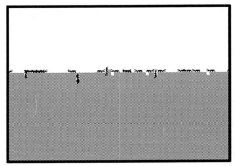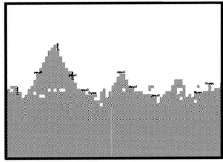

Figure 2-6. Emergent organisation in a simulated ant colony. The pictures show a cross-section of a landscape, with the sky shown in white and the earth in grey. The ants (black figures) wander around the landscape and pick up any loose sand particles they find. The activity of the ants transforms the initially flat landscapes (left) into a system of hills, valleys and tunnels (right) (after Poskanzer 1991).

These artificial life studies of bumblebees, ants and other systems have shown that intricate forms of order can emerge from relatively simple interactions between organisms with each other and with their environment. [8] There is no need for an overall plan. Global organisation is simply a by-product of local interactions.

 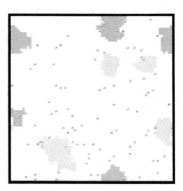

Figure 2-7. Sorting of resources by ants. An initially random collection of objects (left) is rapidly sorted into concentrated piles (right) by ants without any central scheme. The ants move about at random. They pick up objects they find, then drop them again whenever they find another, similar object (Green and Heng 2005).

This idea is extremely powerful. It has implications for many areas of activity. Brooks for instance extended the idea to robotics, where he showed that central intelligence was not necessary for coherent behaviour. In computing, it has encouraged the spread of research on multi-agent systems as a paradigm for addressing complex problems. The ant sort mentioned above, for example, is a simple algorithm that is used for organising incomplete information. Computer scientists are beginning to emulate other aspects of behaviour too. For instance, leaving pheromone trails is now seen as a useful way to get search agents to assist each other in their role of simplifying knowledge discovery on the World Wide Web.

Another form of coordination is seen when groups of animals move through the environment. Some simple kinds of aggregate behaviour emerge because of the way that individuals interact with their environment.

Figure 2-8 shows outputs from a simple model of starfish outbreaks on a coral reef. The interesting behaviour in this case arises from the ways in which the starfish, represented as a set of independent agents, interact with the coral, which is represented as a grid [9]. A crucial property is the percentage of coral cover at each location. At each site, the kind of substrate (reef crest, lagoon, etc.) defines the maximum possible cover, and the coral grows and spreads until it reaches that level. Meanwhile the starfish are agents that can move freely around the reef eating coral (i.e. reducing the amount of cover at each location). The crucial behavioural parameter is the rule that determines where the starfish move. Simple

---

[8] cf. Chapter 9.

[9] We will discuss this kind of model, which is known as a cellular automaton, in the next chapter.

changes to the rule result in completely different patterns in the outbreak. If we assume that the starfish move completely at random, then they quickly spread all over the reef. If we assume that they actively prefer areas of high coral cover, then they spread out along the reef crest, but do not move into other areas.

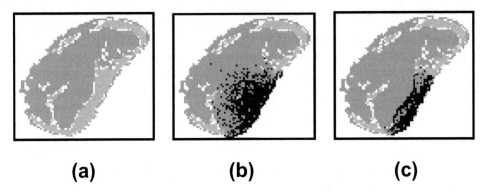

(a)   **(b)**   **(c)**

Figure 2-8. Outputs from the Crown of Thorns (COT) starfish model for several scenarios at Davies Reef, off Townsville Queensland. (a) The underlying reef map is a satellite image of Davies Reef, which has been classified into zones by depth of light penetration. Infested areas are marked in solid black: (b) starfish (dark dots) move around at random; (c) the starfish prefer areas of highest coral cover. See the text for further discussion.

The important lesson to be drawn from the starfish model is that seemingly minor changes in our assumptions about starfish behaviour can lead to enormous changes in the global behaviour of the system. Another insight the model provides is that interactions, whether agent to agent or agent to environment, are important.

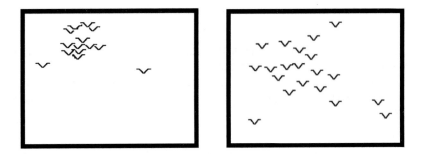

Figure 2-9. Examples of flocking behaviour in a boid simulation of birds.

In many systems, interactions within aggregates of agents, such as those we have just seen, become the main force driving the system. Bees move in swarms, birds in flocks (e.g. Figure 2-9). Fish swim in schools, and zebras move in herds. In a series of simulation studies, Craig Reynolds showed that central coordination was not needed to explain many of the group manoeuvres (e.g. avoiding predators) that biologists have observed (Reynolds 1987). For instance, when attacked by a predator, a school of fish will perform a number of kinds of manoeuvres such as splitting and reforming, as the predator passes through their centre (Figure 2-10). Reynolds' *boids model* showed that observed patterns of this kind could be produced by simple rules governing the behaviour of individuals and the way they interact with other animals nearby. For instance, basic flocking behaviour arises from the following three rules.

    1.   Keep as close as possible to the centre of the group.

    2.   Avoid crowding too close.

    3.   Aim in the same direction as the rest of the group.

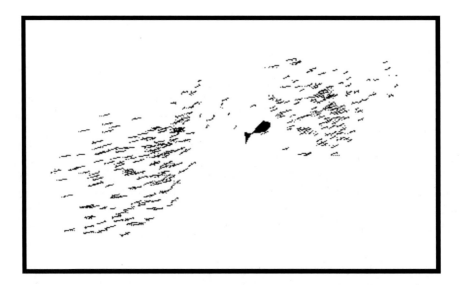

Figure 2-10. Behaviour of a simulated school of fish in the program COOL SCHOOL (Hooper 1999). The behaviour of the school arises because of interactions between the fish, with each individual reacting according to simple rules. Here a school spontaneously divides and regroups to avoid a predator (centre).

Whether it is a flock of birds, a school of fish, or an ant colony, in each of the above examples we see that the whole emerges from interactions amongst the individuals. In later chapters we will see that many kinds of ecological patterns emerge from interactions of one kind or another. In the next chapter, we will see how patterns in landscapes can emerge this way.

CHAPTER 3

COMPLEXITY IN LANDSCAPES

Fire in a forest understorey, southern Australia. In this chapter we will see how the spread of fires such as this depends on the connectivity of fuel across the landscape.

In the previous chapter, we saw that interactions between organisms lead to complexity in ecosystems. Interactions between organisms and their environment also contribute. Soil formation, for instance, involves a combination of organic and inorganic processes. Rainforests not only depend on local climate, but also they modify it. Flow of water across a landscape affects the suitability of sites, determining where plants can grow.

But the story does not end there. Landscapes themselves can be complex. Many biological and physical processes lead to interactions between different parts of a landscape. Plants and soils affect runoff and stream patterns. But the flow of groundwater also shapes the landscape itself. Lakes, hills and streams all affect the movement of plants and animals.

In this chapter, we look at the nature of this complexity that can exist within landscapes, as well as some of its implications.

3.1   THE EYE OF THE BEHOLDER

First things first. Before we can talk sensibly about landscapes and landscape patterns, we must first have some way of representing space, and objects in space. What is more, it is essential to understand the assumptions that underlie ways of modelling landscapes. These assumptions have a bearing on the properties that emerge and on the validity of any conclusions that we might draw from models.

Let's start with location. To indicate any point in space, we use its coordinates. Latitude denotes its deviation (north or south) from the equator; longitude denotes its deviation (east or west) from the Greenwich meridian. Finally, elevation denotes the height of the point above mean sea level. As just indicated, the reference points for the coordinates, where the coordinates take the value of zero, are the equator, the town of Greenwich, England, and mean sea level.

So far so good. Now that we can say where things are in space, we have to be able to indicate what they are. To do this we will adopt the ideas that serve this purpose on computers. The standard way of storing and using data about landscapes is a Geographic Information System (usually called a GIS for short).

3.1.1   Geographic Information Systems

The way we represent landscapes is reflected by the ways that geographic information systems (GIS) store and transform data about landscapes. A GIS organises data about landscapes into layers. A layer is a set of geographically indexed data with a common theme or type. Examples of themes might include coastlines, roads, topography, towns, and public lands. Geographic layers come in three distinct types: vector layers, raster layers, and digital models.

Vector layers consist of objects that are points (e.g. towns, sample sites), lines (e.g. roads, rivers) or polygons (e.g. national or state boundaries). Data of this kind are usually stored in database tables, with each record containing attributes about individual objects in space, including their location.

Raster layers consist of data about sites within a region. Examples include satellite images and digitized aerial photographs. The region is divided into a grid of cells (called pixels when displayed as images), each representing an area of the land surface. The layer contains attributes for each cell. For instance, in a satellite image the attributes might be a set of intensity measurements at different light frequencies, or a classification of the land features within the cell (e.g. forest, farmland, water).

Digital models are functions that compute values for land attributes at any location. For instance, a digital elevation model would interpolate a value for the elevation, based on values obtained by surveying. In practice, digital models are usually converted to vector or raster layers when they are to be displayed or otherwise used.

In the good old days, map layers were literally sheets of clear plastic that had various features drawn on them. Today, they are more likely to be virtual layers that exist inside a computer and become visible only when drawn on a computer screen, or printed out on paper.

There is a big difference between a digital map and a GIS layer. In a digital map, the data are simple coordinates for drawing dots, lines and labels. There may be no connection between the different parts of the data. A river, for instance, might be just a set of disconnected wiggly lines. There may be no indication that they all form part of the same object (the river) and there may be intentional gaps in the lines to allow space to draw roads or to print names. The locations recorded for towns may not be the true locations, but instead indications of where the label is to be plotted. The actual site of a town would be given by a separate dot, with no indication of its link to the label. In contrast, in a GIS, the data are organised into objects. So the record for a river would include both its name and the vector data for drawing it.

The basic tool in GIS is the overlay. That is, we take two or more data layers and lay them on top of one another. To form a map, a GIS user selects a base map (usually a set of key layers, such as coastlines, or roads) and overlays selected layers on top of it.

The process of overlay often involves the construction of a new data layer from existing ones. For instance, if we overlay park boundaries on top of forest distributions, then we can create a new map showing areas of forest within parks.

Both vector and raster representations of plant distributions have advantages and drawbacks. Vector models (which represent individual plants as points in space) can require a great deal of computer time and are therefore difficult to apply to problems on large scale or heterogeneous landscapes. Raster models, in which "pixels" represent areas of the landscape, are less precise but more tractable for computation. They also reflect major sources of large-scale data, such as satellite imagery, which consists of arrays of pixels.

3.1.2   Cellular automata models of landscapes

Now that we have a way of representing things in the landscape, we need ways of using that data to simulate the processes that go on there. Each of the three types of data layers that we met above are sometimes used for modelling. However, models based on raster data are perhaps the simplest and most common. Here, we look at ways in which models of landscapes are built using raster data. In Chapter 9 we shall come back to the question of modelling with vector data when we look at agent models, and in Chapter 10 we shall look in more detail at the technology that underlies the modern development and use of geographic information, as well as the modelling tools that make use of them.

In a raster-based model, we represent the landscape as a rectangular array of cells (Figure 3-1). Each cell contains information about the area that it represents. The figure shows an example in which the cells are shaded by landscape category, including fields, trees or roads. In this example, the amount of fuel found at each site is given by a separate table, which lists the amount of fuel associated with each category. However, in a model of this kind, the cells could contain many other kinds of data, such as vegetation type, altitude, slope, geology.

Raster models generally use the formalism of cellular automata to simulate plants in landscapes. A cellular automaton (Wolfram 1984) is an array of interacting cells, with each cell behaving according to the same program. The essential features of a cellular automaton ("CA" for short) are as follows:

- its state is a variable that takes a separate value for each cell;
- its neighbourhood function defines the set of nearby cells that each cell interacts with. In a grid, the neighbourhood normally consists of cells physically closest to the cell in question (see Figure 3-2);
- its program is the set of rules that define how each cell's state changes in response to its current state, and those of its neighbours.

Cellular automata models of landscapes consist of fixed arrays in which each cell represents an area of the land surface (Green 1989). The states associated with

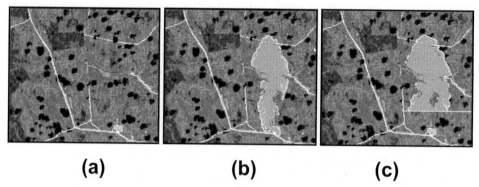

(a)   **(b)**   **(c)**

Figure 3-1. A raster-based model of fire spread across a landscape. (a) The landscape map, taken from a digitized aerial photograph, is a rectangular grid of cells that are here coloured to indicate their contents. The white lines are roads; the black spots are trees, and the shaded areas represent different kinds of land cover, mostly fields and grasslands. The spot near the bottom (centre) represents a house. (b) A scenario in which a fire starts from a cigarette butt thrown from a car under hot, dry weather and driven by a strong northerly wind. The house has just been destroyed. The uniform grey area indicates locations that have been burnt out, and the bright area surrounding the grey area shows the fire front. (c) Same scenario but testing the effectiveness of cutting a break across the fire's path.

each cell correspond to environmental features, such as coral cover or topography. This approach is compatible with both pixel-based satellite imagery and with quadrat-based field observations. It also makes it relatively easy to model processes, such as dispersal or fire (Green et al. 1990), in heterogeneous landscapes (Figure 3-1).

The most important feature of the CA model is the role of the neighbourhood (Figure 3-2). Any cell, taken in isolation, behaves in a certain simple way, just like a tree that grows by itself in a glasshouse will grow. However, just as trees in a forest interact with each other to produce a rich variety of growth, form and dynamics, so it is the interactions of each cell with its neighbours that dominate the behaviour of a cellular automaton model.

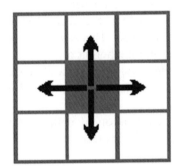 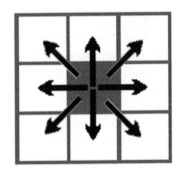

Figure 3-2. Examples of neighbourhoods in two different cellular automata (CA). The two diagrams shown represent a typical cell (shaded) in a CA grid. The blue arrows indicate the nearby cells that form part of the shaded cell's neighbourhood. As shown by the arrows, in the CA at left, each cell has 4 neighbours; in the CA at right, each cell has 8 neighbours.

The first and most famous CA model is the game "LIFE". Invented by the Cambridge mathematician John Conway, LIFE is a simple 2-D analogue of basic processes in living systems. The game consists in tracing changes through time in the patterns formed by sets of "living" cells arranged in a 2-D grid. The behaviour of LIFE is typical of the way in which many cellular automata reproduce features of living systems. That is, regularities in the model tend to produce order. Starting from an arbitrary initial configuration, order usually emerges fairly quickly in the model. This order takes the form of areas with well-defined patterns. Ultimately most configurations either disappear entirely or break up into isolated patterns. These patterns are either static or else cycle between several different forms with a fixed period. (See Chapter 4 for a more detailed discussion and examples.)

Notice that CA models adopt the raster representation of landscapes, which we introduced in the previous section. The alternative vector representation is embodied in what are known as agent-based models. One important difference between the two kinds of model is a matter of scale. By definition, individual-based models act at the scale of individual plants and animals. In contrast, the states in a CA model summarize the contents of an entire area. Both the size of the area and the detail is variable. Even if they are set up to simulate exactly the same process, cellular automata models and individual (agent based) models of (say) forest dynamics can produce somewhat different results (Lett et al. 1999).

Ecologists have applied cellular automata models to many problems in landscape ecology. One common application has been to identify the way in which particular kinds of spatial patterns form. For example, a team of Argentinean scientists used CA models to look at the vegetation patterns that resulted from the combination of tree growth rates and the killing capacity of the wind in the subantarctic forest of Tierra del Fuego (Puigdefabregas et al. 1999). They were able to show that simulated patterns for heterogeneous forests with random age distributions matched the patterns observed in nature.

Similarly, CA models have been used to examine the formation of wave patterns in the heights of trees (Sato and Iwasa 1993). This study, by Sato and Iwasa, found that both "absolute height (or age) and the height difference between neighbours affect tree mortality". Another study used CA models to examine interactions and vegetation degradation on a contamination gradient (Walsworth and King 1999).

We shall see many more examples of the application of cellular automaton models to ecological questions, both below and in the chapters that follow.

3.2   SAMPLING AND SCALE

What is the point of sampling? A simple question, and yet it is one that is all too easy to forget. Collecting data and samples from different locations in a landscape is one of the most basic jobs in ecology. However, ecologists often resort to some standard pattern of sampling without considering whether it really will answer the crucial questions.

In landscapes, the patterns that we see are reflections of the processes that produced them. Dry hilltops and lush valleys reflect the movement of water and soils. The distributions of plants and animals arise from a multitude of processes that we have to tease apart. Sampling methods stem from these assumptions. Now as we have seen, a lot of traditional theory in ecology is based on reductionist models of cause and effect, and often this assumption is correct. Plant A grows at site X because of soil type Y.

In keeping with the two main kinds of geographic data layers (vector and raster) introduced in the previous section, there are two ways of collecting spatial data. We can record objects and note their location (vector data), or we can record locations, and note their features (raster data). Both approaches are used in ecology, but the latter is more common in field studies. Ecologists have normally used vector data to map large-scale features, such as lakes and streams, or areas of different vegetation type, such as swamp, grassland or forest. They also use vector data to record animal territories. Vector (point) data have sometimes been recorded during intensive studies of spatial pattern, such as mapping every tree within part of a forest.

The most common methods of sampling landscapes are quadrats and transects. A *quadrat* is a square within which an ecologist takes samples or records observations. She might, for instance, record every plant within the quadrat, or estimate biomass. The size of quadrats varies according to the degree of spatial variation within the landscape. Metre square quadrats are common in field ecology, but in specialised studies they may vary in size from less than a millimetre to hundreds of kilometres. A *transect* is a set of quadrats in sequence across a landscape. Transects are often used to sample some quantity that is known to vary systematically across a landscape, such as elevation, water table depth or nutrient levels..

The usual intent of quadrat sampling is to obtain a sample that is representative of an entire landscape. To allow for random variations, lots of quadrats may be needed. Transects are a way of taking into account systematic variations, such as differences in soil moisture down a slope. In essence, these methods cater for reductionist interpretations. They assume uniformity and attempt to eliminate known sources of non-uniformity. However, it may be that non-uniformity itself is the most telling pattern of all.

The essential problem of landscape complexity is that as often as not, complex processes leave behind complex patterns. To interpret a complex pattern often requires a lot of data. You may have to map features in detail, and on different scales. Sometimes it is simply not feasible to gather detailed information about an entire landscape. Quadrat samples are a compromise between what is desirable and what is possible. They allow field workers to put together a picture of an entire landscape at the cost of reasonable time and effort. Fortunately, new technologies, especially airborne or satellite remote sensing, are making it possible to map entire landscapes, even if some of the fine scale detail may be lost. As we shall see later in this chapter, methods are now available to interpret aspects of the complexity that we see in landscape patterns.

3.3   COMPLEXITY IN SPATIAL PROCESSES

Spatial processes are inherently complex. Water percolates through the soil. Wind carries seeds to fresh territory. Chestnut blight spreads from tree to tree. In almost every case, spatial processes involve interactions between objects at different locations in a landscape.

The patterns that we see in landscapes are often like frozen memories of the past. Fly over the landscape of eastern Canada, Finland, or Russia, and what strikes you the most are the lakes. There are thousands of them: reminders of the distant Ice Age, they break up the landscape and have immense influence on the ecology of those regions. Likewise, if you fly over the scrublands of Western Australia, you see striations in the vegetation, which are the results of fires burning parallel to the sand dunes.

To understand patterns such as these, we can model the processes that lead to them. One important class of processes is percolation. Percolation involves

decreases, the estimated length increases without limit. Thus, if the scale of the (hypothetical) measurements were to be infinitely small, then the estimated length would become infinitely large!

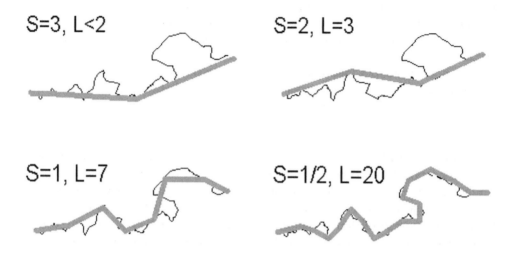

Figure 3-3. Using sticks of different size S to estimate the length L of a coastline.

When discussing measurement, *scale* can be characterized in terms of a measuring stick of a particular length: the finer the scale, the shorter the stick. Thus at any particular scale, we can think of a curve as being represented by a sequence of sticks (Figure 3-3), all of the appropriate length, joined end-to-end. Clearly, any feature shorter than the stick will vanish from a map constructed in this way. Of course, no one actually makes maps by laying sticks on the ground, but the stick analogy reflects the sorts of distortions that are inevitably produced by the limited resolution of aerial photographs, by the length of survey transects, or by the thickness of the lines produced by drafting pens. There is an analogy here, too, with the accuracy or frequency with which any sort of biological measurements are made.

The dependence of length (or area) measurements on scale poses serious problems for biologists who need to use the results. For example, lakes that have a very convoluted shoreline are known to offer a larger area of shallows in relation to their total surface area, and thus support richer communities of plant and animal life. Attempts to characterize shore-line communities in terms of indexes that relate water surface to shoreline length have been frustrated by problems of scale.

Mandelbrot proposed the idea of a fractal (short for "fractional dimension") as a way to cope with problems of scale in the real world. He defined a fractal as any curve or surface that is independent of scale. This property, referred to as self-similarity, means that any portion of the curve, if blown up in scale, would appear identical to the whole curve. Thus the transition from one scale to another can be represented as iterations of a scaling process (e.g. see Figure 3-3).

In the coastline example, the implication of Mandelbrot's definition is that as the scale of your measurements decreases, the total distance that you measure increases. So hypothetically, the answer to the question we asked at the beginning is that the coastline of Britain would be infinite in length!

An important difference between fractal curves and the idealised curves that are normally applied to natural processes is that fractals are rough. Although they are continuous - they have no breaks - they are "kinked" everywhere. We can characterize fractals by the way in which the representation of their structure changes with changing scale.

3.4.1 Fractal dimensions

The notion of "fractional dimension" provides a way to measure how rough fractal curves are. We normally consider lines to have a dimension of 1, surfaces a dimension of 2 and solids a dimension of 3. However, a rough curve (say) wanders around on a surface; in the extreme it may be so rough that it effectively fills the surface on which it lies. Very convoluted surfaces, such as a tree's foliage or the internal surfaces of lungs, may effectively be three-dimensional structures. We can therefore think of roughness as an increase in dimension: a rough curve has a dimension between 1 and 2, and a rough surface has a dimension somewhere between 2 and 3. The dimension of a fractal curve is a number that characterizes the way in which the measured length between given points increases as scale decreases. Whilst the topological dimension of a line is always 1 and that of a surface always 2, the fractal dimension (D) may be any real number between 1 and 2.[1]

For the shoreline shown in Figure 3-3, the fractal dimension of the coastline is about 1.5. This is a measure of how crinkly it is. A perfectly straight line would have a dimension of 1. At the other extreme, a curve of fractal dimension 2 would be so convoluted that it would fill all of two-dimensional space.

Returning to the L-system representations, described in Chapter 2, for plant like structures, the production rule: A → FA will have a fractal dimension of 1 and as the order of branching increases (e.g. N → F[-FL][+FR]FA and N → F[-FA][+FA]FA and so on) the fractal dimension will approach 2 for 2-dimensional patterns.

---

[1] The fractal dimension D is defined by the formula $D = \dfrac{\log(L_2/L_1)}{\log(S_1/S_2)}$ where L_1, L_2 are the measured lengths of the curves (in units), and S_1, S_2 are the sizes of the units (i.e. the scales) used in the measurements. For the shoreline shown in Figure 3-3, measurements for $S_1=1$ and $S_2=1/2$ give lengths of $L_1=7$ and $L_2=20$, respectively. So entering these numbers in the calculation gives the result $D = \dfrac{\log(20/7)}{\log(2/1)} = 1.51.$ [0]

Now the idea of fractals is built on the assumption that patterns repeat at different scales. But in the real world, this is not necessarily true. Different processes influence patterns on different spatial scales. In this case, if we repeat the calculation for the transition from $S=1$ to $S=2$, then we find the smaller estimate of $D=1.22$; and the transition from $S=2$ to $S=3$ gives $D \sim 1.13$. In other words, the coastline becomes smoother at larger scales.

It is important to realise that the above way of estimating fractal dimension applies only to certain sorts of data. Suppose that we wish to measure fractal properties of, say, the surface of a coral reef. There are two different measurements that we might make. One measurement would consist of measuring distances between two points on the reef with measuring sticks of varying lengths (as in Figure 3-3). If instead we moved along the same transect and measured, say, the height of the reef surface above the substrate, then we could not measure fractal index in the same way.

No curve or surface in the real world is a true fractal; they are produced by processes that act over a finite range of scales only. Thus estimates of D may vary with scale, as they do in the above example. The variation can serve to characterize the relative importance of different processes at particular scales. Mandelbrot called the breaks between scales dominated by different processes "transition zones".

Fractal geometry has become important in many fields. As we shall see in later chapters, it is related to several important ideas about complexity, such as chaos (see Chapter 4). The repeating nature of fractal patterns is intimately related to basic computation, which consists of repeating operations. It also ties fractals closely to iterative processes in nature, such as cell division. Branching patterns arising during plant growth, for example, are inherently fractal in nature, as we saw in Chapter 2.

3.4.2   Fractals in nature

Many natural structures have fractal properties. In biological patterns the fractal nature arises from the iteration of growth processes such as cell division and branching. Structures with high fractal dimension, such as lungs and branches, are associated with processes that maximise surface to volume ratios.

Fractals in nature arise from the action of specific processes. One of the useful insights to be gained from fractals is to help us understand the roughness that we often see in natural things. Landscapes, for instance, are often rough at all spatial scales. This roughness is a result of natural processes, such as weathering, that operate on many different spatial scales. Fractal models capture that roughness.

Unlike theoretical models, these processes operate only over a finite range of scales. For this reason the fractal dimension of many natural structures remains constant only over a limited range. Sometimes there are distinct breaks between scales, where one process ceases to become important and another becomes dominant. In coral reefs, for example, coral growth, buttress formation and underlying geomorphology all affect the profile. Each process operates on different scales and the fractal dimension of the reef surface has distinct breaks at the transition from one process to another.

Fractal dimension provides a way of measuring the complexity of landscape patterns (e.g. Turner 1989). One application, for instance, has been to examine some of the fractal properties that arise in rainforest gap analysis (Manrubia and Solé 1996, 1997). For instance, a region with a high fractal index will have a long, convoluted boundary relative to its area. So edge effects are likely to be a major consideration for wildlife conservation in such a region. The increasing use of geographic information systems and satellite imagery to study landscapes and landscape patterns has helped to popularize this approach.

3.5   ARE LANDSCAPES CONNECTED?

We can define a set of sites in a landscape as connected if there is some process that provides a sequence of links from any one site to any other site in the set. In the CA formalism, connectivity is defined by the neighbourhood function. Two sites are directly connected if one belongs to the neighbourhood of the other. A region in a landscape is connected if we can link any pair of points in the region by some sequence of points (i.e. a path or "stepping stones") in which each pair of points is directly connected.

Before we go any further, let us be clear about what exactly we mean by the word "connected" in ecological terms. Two objects are "connected" if some pattern or process links them. Dingoes, for instance, are linked to kangaroos because they prey on them. The most important sources of connectivity for plants and animals are associated with landscapes. Links within a landscape arise either from static patterns (e.g. landforms, soil distributions, or contiguous forest cover) or from dynamic processes (e.g. dispersal or fire). For instance, a tree growing within a forest can transmit and receive pollen from any other tree that lies within the range of bees or other pollinating vectors. Pollination provides a connection among the trees.

We need to make clear several basic aspects of connectivity. First, it is important to realise that a landscape may be connected with respect to one process, but not with respect to another. A river or fence running through the middle of a forest may form a barrier to restrict some processes (e.g. fire or movement of animals) but not others (e.g. movement of bees or wind-borne seeds). Secondly, we need to distinguish between the connectivity of any two objects (e.g. two trees) and the connectivity of an entire system. For instance, a tree at one end of a forest may not be able to pass pollen directly to trees at the other end of the forest, but over a period of generations its genetic information can flow throughout the forest.

In a patchy environment, when is a landscape connected and when is it not? To answer this question, we can simulate a patchy landscape by using a grid in which each cell represents an area of the land surface (Green 1989, 1994a, 1994b). We classify each cell according to (say) the dominant vegetation within the area it represents. Now suppose that there is a random distribution of (say) rainforest within the region represented by the grid. Let us assume that two cells containing rainforest are connected if they are adjacent to one another. In biological terms this assumption means that some process that we are interested in (e.g. dispersal, animal migration) links adjacent cells. Let us define a patch of rainforest as a set of rainforest covered cells in which we can move anywhere within the patch via a sequence of adjacent cells. In this system we find that the connectivity of sites in a landscape falls into three distinct phases: disconnected, critical, and connected (Figure 3-4). If the density of rainforest sites is sub-critical, then the landscape is broken into many isolated sites and small patches. If the density is critical then a single large region may be connected (black shading in Figure 3-4), but much of the landscape remains as small isolated patches. If the density is super-critical, then almost the entire landscape is connected, with few isolated sites remaining (Figure 3-4).

25%   **50%**   **75%**

Figure 3-4. Density and connectivity of sites in a landscape. The numbers indicate the proportion of covered sites (represented by black coloured cells) in the region. Notice that in lower percent cover, connected patches are few, but as the amount of cover increases, they join to form patches. When 75% of the cells are covered, the covered sites form a connected patch that encompasses the entire area.

The relationship of the above results to other kinds of criticality (Bak and Chen 1991) and to percolation theory is well known (Stauffer 1979, Wilkinson and Willemsen 1988). As the name implies, percolation is to do with flows through a medium. As we saw above, the ability of a flow to spread through a medium depends on the formation of "edges" within a lattice, and is usually determined by density. A phase change occurs when a critical density is reached. More recently it has been shown (Green 1993, 1994a) that all of these criticality phenomena stem from underlying properties of graphs (sets of nodes and edges). Perhaps the most striking implications of the above phase changes concern epidemic processes, as in the following examples.

• Fire spread depends on the density of fuel being greater than a critical level; otherwise the fire quickly dies out (Green et al. 1991). The critical density is a function of environmental factors such as temperature and fuel moisture. This phenomenon has been confirmed in both laboratory experiments and field studies[2].

• Simulation studies of the spread of Crown of Thorns starfish outbreaks on the Great Barrier Reef suggest that reef-to-reef spread relies on inter-reef connectivity, as defined by the dispersal of starfish larvae between reefs being above a critical threshold (Bradbury et al. 1990).

• The invasion of a new species in a region depends on there being a critical density of sites available for colonisation (Green 1990). Many weeds and pests are exotic species that perform best in cleared areas. Therefore the connectivity of cleared areas is as important a consideration as the connectivity of undisturbed habitats. These findings are currently being applied to determine control scenarios in the spread of exotic diseases such as foot and mouth disease (Pech et al. 1992).

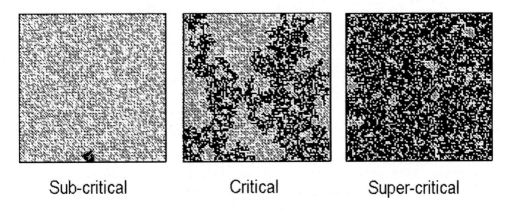

Sub-critical   Critical   Super-critical

Figure 3-5. Phase change in the connectivity of a cellular automaton grid as the proportion of "active" cells increases (after Green 1994a). Grey denotes active cells; white denotes inactive cells. The proportion of active cells increases from left to right. The black areas indicate connected patches of active cells. Notice that a small change in the number of active cells produces a phase change in the system from many small patches, isolated from one another, to essentially complete connectivity of the entire system.

The key result to emerge from studies of connectivity (Green 1994a) is that landscapes can exist in two different phases: connected and disconnected (Figures 3-5, 3-6). The variability that occurs at the phase change (Figure 3-6b) means that the size and distribution of landscape patches become highly unpredictable when the density of active regions is at the critical level.

---

[2] Laboratory studies include Beer and Enting (1990), Nahmias et al. (2000) and field studies include Caldarelli et al. (2001).

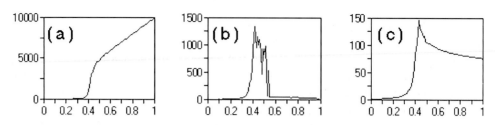

Proportion of cells that are active

Figure 3-6. Critical changes in connectivity of a CA grid (compare Figures 3-5, 3-6) as the proportion of active cells increases (x-axes). (a) Average size of the largest connected subregion (LCS). (b) Variation in the size of the LCS. (c) Traversal time for the LCS. Each point is the result of 100 iterations of a simulation in which a given proportion of cells (in a square grid of 10,000 cells) are marked as active. Note that the location of the phase change (here ~0.4) varies according to the way we define connectivity within the model grid.

3.5.1 *Why is a starfish like an atomic bomb?*

On November 5th, 1983, the Weekend Australian carried the following startling headline:

"Reef park a farce in face of starfish threat".

This item, along with other sensationalist reporting, sparked a frenzied debate (Raymond 1986), followed by intense research, into the problem of starfish infestations along Australia's Great Barrier Reef. The controversy erupted because the Crown of Thorns starfish (*Acanthaster planci*) was appearing in unprecedented numbers on several reefs. In reef after reef, the entire coral population was being literally eaten away while distraught tour operators could only look on in horror.

It took four years of research by ecologists at the Institute of Marine Science in Townsville to uncover what was really going on. The team, led by Roger Bradbury, first identified the pattern of outbreaks. The Great Barrier Reef is not a single reef, but a chain consisting of thousands of individual coral reefs. It runs parallel to the North Queensland coast for over 2000 kilometres. The team began by surveying hundreds of reefs. Destruction was patchy. Some reefs were totally denuded by the starfish; others nearby were completely untouched.

A plot of the outbreaks revealed a travelling wave of outbreaks that moved from north to south. This wave pattern was weakest in the north, where outbreaks seemed to be more sporadic, but became progressively stronger as it moved south. The systematic pattern implied a causal relationship. Further work revealed that offshore ocean currents carried larvae from one reef to the next. The starfish give birth to their young as larvae on a single night each year. The sea becomes filled with clouds of tiny larvae. These get caught by ocean currents which carry them to other reefs, where they settle in huge numbers. Then the cycle starts again.

In atomic fission, a neutron hits a nucleus and splits it. In the process, more neutrons are released. These neutrons then split other nuclei and the process continues in a chain reaction. The effect of starfish on the Great Barrier Reef is very similar. Larvae settle on a reef, setting off a local outbreak. When they become adults, they release more larvae, which currents carry to the next reef. In this way, they set off a chain reaction of outbreaks.

In a simulation model of this process (van der Laan and Hogeweg 1992), the ecologists combined data on currents with a map of the entire Great Barrier Reef. This model duplicated the patterns revealed by surveys and confirmed that the "epidemic" model was indeed correct.

The only remaining question was how the outbreaks started in the first place. Obviously the process has to be seeded by outbreaks in the waters north of Cape York, but how? Without massive influxes of starfish larvae, why should an outbreak suddenly start? Is it chance? Or is it something more serious? The most contentious theory is that over-fishing reduces the predators that normally feed on starfish larvae and young starfish. The issue is still hotly debated to this day.

The Crown of Thorns starfish provides an excellent example of biological connectivity. The starfish larvae move from one reef to another, setting off outbreaks wherever they settle. More generally, we can consider landscape connectivity in terms of the plants and animals that inhabit it. Many animals, for instance, have a home range or territory within which they roam. That territory defines a relationship that connects every site within it.

This kind of analysis is most useful where animals inhabit isolated or fragmented habitats. The assumption is that two habitats are connected if an animal can reasonably move between them. In practical terms, most researchers have considered the distance that an animal might normally move to get from A to B. Long distance transport, migration and freak events are not considered.

Rodney van der Ree used this approach to study the distribution of squirrel gliders in fragmented forests that consisted mostly of roadside strips bordering farmlands (van der Ree 2003). The question was whether the gliders would reach isolated trees and patches that were separated by open fields from the roadside cover that formed their chief habitat. He found evidence for a threshold of 75 metres. That is, the incidence of gliders visiting trees further than this distance in the course of their normal activity was very small.

Some birds can fly great distances in the course of a single day. This can lead to landscape connectivity on a grand scale. For instance, David Roshier and his colleagues looked at the connectivity of water bodies in central Australia for the water birds that inhabit them (Roshier et al. 2001). They did this by assuming that water birds would not normally fly more than about 200 kilometres in the course of a day's travel while moving from one lake or swamp to another (cf. Chapter 8). On this basis, they found that during wet years the entire continent was effectively connected.

It is important to realise that habitat connectivity will vary from species to species. Just because one species finds a habitat connected does not mean that this is the case for all species. For example, Joanne McLure showed that in the

Apennine Mountains of central Italy, the environment is well connected for wolves but not for less mobile species such as bears (McLure 2003).

These differences between species can influence species richness. In Brazil's Sao Paulo state, Jean Paul Metzger studied the effect of the area and connectivity of forest fragments on the diversity of trees and vertebrates (Metzger 2003). He found that forest cover was important at local scales, but, in terms of the whole landscape, connectivity between fragments was a better indicator of species richness.

The above studies all considered connectivity patterns evident in the environment. However, populations may be fragmented even in the absence of corresponding landscape patterns. Similarly, environmental conditions may change, so populations may have been fragmented in the past. We will look further at some implications in later chapters.

result of many individual birds flying together. An ant colony is the product of the activity of many individual ants.

Three aspects of aggregates of individuals influence the nature of the system that emerges. First, there is the character of the agents themselves. Bees in a swarm behave differently from birds in a flock. The second aspect is the quality of the interactions between the agents. Individuals in a rioting mob interact very differently from guests at a cocktail party. The third aspect is the "wiring pattern" in the network of interactions between agents. The pathways of influence in a feudal kingdom are different from those in a democracy.

An important issue is whether the interactions persist across different scales. In a gas, for instance, the main interactions (molecules colliding and rebounding off one another) are all brief, local and linear. They quickly average out so that the nature of the gas remains the same over a wide range of scales. The temperature of the gas is an emergent property (see Chapter 1) that expresses the average behaviour. In contrast, non-linear systems do not usually scale in such a simple fashion. They are governed by non-linear interactions, which are a common source of complexity.

In physics, entropy is a measure of the disorder, or randomness, in a system. For example, a system with hot spots and cool spots has order: there are patterns of heat and cold. A system in which the temperature is uniform throughout has low order, and high entropy. The Second Law of Thermodynamics states that in any closed system, the entropy of the system will increase with time until the

Figure 4-4. A forest emerges out of the interactions between millions of trees and plants, both with each other and with the physical environment.

system is uniformly disordered. For instance, if you heat the end of a metal bar, the heat will spread until the whole bar becomes hot. An ice cube melts in an oven. In both cases, order decreases.

One of the puzzles of physics was to account for living systems, which seem to fly in the face of thermodynamics by accumulating order. Likewise, why do clouds of interstellar gas condense into stars? According to the second law of thermodynamics, this should not happen. But the second law applies explicitly to closed systems, whereas living systems are open. They exchange energy and materials with their environment. Therefore, order can increase locally in gas clouds and in living systems because of the exchange of energy with the outside.

The Nobel prize winner Ilya Prigogine developed the idea of *dissipative systems* to explain how some systems can accumulate order. Dissipative systems are open systems that are maintained in an orderly state by exchanging energy with their environment. That is, they take in energy from their environment, use that energy to generate orderly internal structures, and dissipate it to the environment in a less orderly form.

In dissipative systems, there is no tendency to smooth out irregularities and for the system to become homogeneous. Instead, irregularities in dissipative systems can grow and spread. Chemical systems provide many examples, such as crystal formation. However, by far the most common dissipative systems are living things.

Within intergalactic dust and gas clouds, gravitational interactions between particles lead to the formation of stars and planets. Unlike gas in a bottle, in which molecules rebounding off one another serve to distribute energy evenly throughout the system, gravity is the predominant interaction within interstellar clouds. The global properties of interstellar clouds are therefore very different from bottled gas. Instead of moving towards homogeneity, irregularities in these clouds coalesce into stars, planets and other astronomical objects.

4.3.2 Modularity

Above we saw that pattern and organisation can arise out of the interactions between things. However, complex structures do not magically appear. One of the most important principles at work is modularity.

To understand modularity, a good analogy is a carry bag. Think of any bag that you regularly use to carry things, such as a hiking pack, a purse, or a briefcase. The simplest carry bag is just a sack. You throw all the items that you want to carry into the sack and off you go. If there are just one or two, then there is no problem. But people often need to carry dozens of separate items around with them. If you throw them all into a sack, then you could be rummaging around for ages trying to find what you want. Even worse, some items might be incompatible. Your lunch could leak all over your wallet, for instance.

So what do we do? We create modules. Hikers, for instance, put clothes in one bag, cooking gear in another bag, and so on. In this way, different classes of items are kept together. This makes it easy to find the cooking gear when you need it, and avoids dirty pots from soiling your clean clothes. Most purses and briefcases provide pockets and compartments to help us store the contents in modular fashion.

The same principle holds in many different systems. Large organisations divide their business into sections, branches, divisions and so on. This simplifies the way they operate and enables people to cope. Engineers use modularity too. Any large system, such as an aircraft or a factory, may consist of thousands or even millions of individual parts. A common source of system failure is undesirable side effects of internal interactions. The problem grows exponentially with the number of parts, and can be virtually impossible to anticipate.

The solution is to organise large systems into discrete subsystems (modules) and to limit the potential for interactions between the subsystems. This modularity not only reduces the potential for unplanned interactions, but also simplifies system development and maintenance.

The same is true of cities. To support huge concentrations of people, cities have to provide a wide range of services, including: housing, transport, distribution of food and other commodities, water, sewage, waste disposal, power, and communications. On top of these vital services, there are social infrastructures such as education, police, fire brigade, hospitals, ambulance, and shopping centres. The interactions of so many systems, combined with rapid growth, technological development, and social change, underlie many problems in modern society.

The living world around us is also *modular*. It is full of hierarchies. The body plans of all organisms are highly modular. The human body, for instance, consists of discrete organs, bones, and tissues. Organs form bodies. Each organ is built out of discrete cells. Likewise, the cells are modular too. They each consist of distinct organelles: nucleus, mitochondria, and so on. As we shall see in the next chapter, plants are even more modular than animals. A tree, for instance, consists of thousands of nearly identical branches, leaves, buds, flowers and fruit.

At a higher level, species themselves can be seen as a form of genetic modularity. They represent reproductively closed populations. Ecosystems too are modular. They consist of distinct populations, and may also contain niches or habitats that are partially closed subsystems. Likewise, different parts of a landscape may form modules. Interactions within a pond or stream, for instance, are often much richer than with the surrounding woods.

Although hierarchies reduce complexity, as described above, they also introduce brittleness into a system. Removing a single node, or cutting a single connection breaks the network into two separate parts. Every node below the break becomes cut off. This brittleness occurs because hierarchies are minimally connected. There is no redundancy in the connections. For instance, the Internet is organised as a hierarchy of domains. If a domain name server fails, then computers in that domain may be cut off from the Internet. Centralised services, such as city power supplies, suffer from the same problem.

Recent studies suggest that instead of hierarchies, many large systems, including metabolic pathways, large societies, and some ecosystems, are organised as scale-free networks (Figure 4-3f). These are networks in which most nodes have few edges linking them to other nodes, but some have large numbers of connections to other nodes. In contrast, in a simple hierarchy, such as a binary tree (Figure 4-3d), most nodes have three connections per node (one up and two down). Scale-free networks have two important features. One is that they are independent of scale; remove any number of nodes and the overall structure still looks much the same. The other is that they are brittle in face of systematic disturbances. Removing a few critical nodes from a tree breaks the system apart. However, a scale-free network can retain full connectivity even when large numbers of nodes are removed.

As we have seen here, complex networks pervade our world and offer a powerful way to look at a wide range of processes. Understanding the structures that occur in networks, and how these influence processes within them, can provide insights into the functioning of complex systems. These structures include trees, cycles, hierarchies, and modules. Some key processes are positive and negative feedback, percolation, phase transitions, criticality and self-organisation. Network theory is closely tied to dynamical systems and chaos theory. In the next chapter, we shall look at these ideas and how they relate to some major questions in ecology.

CHAPTER 5

THE IMBALANCE OF NATURE

A peaceful lakeside scene such as this may seem unchanging, but over long periods the forests can undergo massive change. Shown here is Everitt Lake, Nova Scotia, which is discussed in Chapter 6.

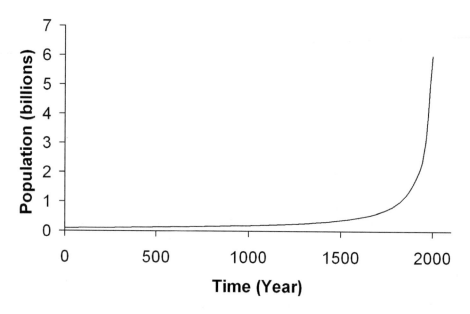

Figure 5-1. Exponential growth in the world's human population. This curve shows the historical trend in the growth of the world's human population (in billions) over the last two thousand years.

population, which may drop more quickly than linearly in response to increasing use.

In any of these scenarios, the population will fluctuate. As the population grows, resources become severely and patchily depleted. Many individuals die, but not before they have further depleted resources that the survivors require. Thus, the population can drop rapidly and suddenly for no obvious reason. We term this a population crash. In economics, fluctuations of this kind are known as booms and busts.

One of the biggest factors that loosens the connection between population growth and resource availability is seasonal variation. This causes the fluctuations depicted in the right panel Figure 5-2. If breeding occurs in spring when resources are abundant, but food becomes scarce in winter, then populations can grow unsustainably in spring, only to suffer winter famine. This, incidentally, is a major reason why populations in the tropics are more stable and less vulnerable to extinction than populations in the Arctic, as we mentioned in Chapter 1. It also results in migration for some species.

A severe crash can drive a small population to extinction. For this reason, ecologists and mathematicians are intensely interested in the causes of fluctuating population sizes. Unfortunately, one of their major findings is that both the timing of population booms and crashes, and the size of these fluctuations, are usually unpredictable.

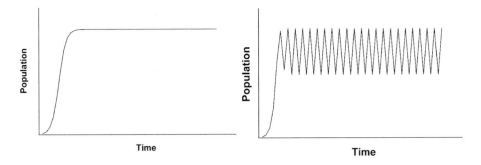

Figure 5-2. Logistic growth for populations that reproduce continually (left) and seasonally (right). The cycles occur because the population size repeatedly jumps above the carrying capacity, which initiates a population crash.

There are two reasons for this unpredictability of population booms and crashes. First, as mentioned above, there may be large, effectively random factors that alter the carrying capacity or the growth rate from year to year, such as climate variation and interactions with other species. Secondly, if the population growth rate is high, then the logistic equation falls into a state in which even the smallest random disturbance can change its entire future behaviour. This state is called deterministic chaos.

Later in this chapter we will look in detail at what deterministic chaos is, why it occurs, and how it affects predictions about populations. Also, we have not considered the additional complexity generated by migration within and between populations. We will look at that issue in Chapter 6. First, however, we need to recast these ideas in more general terms by looking at populations as dynamic systems.

5.2   WHO EATS WHOM?

How do we describe the changes that go on in a system? Often we can express important properties, such as population size or annual profit, in terms of numbers. So changes within the system are reflected by the changes in the values of the *variables* that represent these properties.

Changes through time are known as the *dynamics* of a system. A set of interrelated variables (for example, population size) whose values change through time is called a *dynamic system*. Besides the variables used to represent values of particular quantities, dynamic systems may also include variables to represent rates of change through time. For example, in a predator-prey system, such as foxes eating rabbits, we would have variables for the size and reproduction rate of each population (Figure 5-3). So the dynamic system would have four variables:

- size of the rabbit population;
- size of the fox population;
- rate of change for the rabbit population;
- rate of change for the fox population.

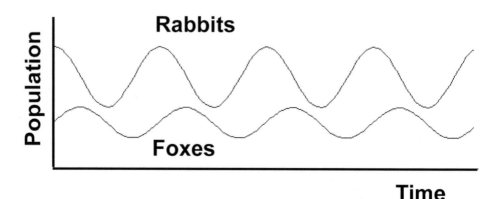

Figure 5-3. Population changes in a simple predator-prey system.

Besides system variables, a dynamic system also includes *parameters*. These are fixed values that determine particular properties and relationships. For instance, in the case of the foxes and rabbits system, we have four parameters:

- birth rate (per season) for rabbits;
- birth rate (per season) for foxes;
- mortality rate for rabbits;
- mortality rate for foxes.

To fully define this dynamic system we also need to specify the *initial conditions*, that is the values of all the variables (in this case, the sizes of the populations) at the start.

We can represent the behaviour of a dynamic system in several ways. The most obvious is to plot values of the variables (vertical axis) versus time (horizontal axis), as shown in Figures 5-3. Where there is more than one variable involved, an alternative way to view change is to plot one variable against another to show the trajectory that the system follows. For instance, for the foxes and rabbits system (Figure 5-4) each point on the graph defines a state of the system. So the area on the graph shows all possible states of the system. This is called the *state space* of the system. The trajectory shown in Figure 5-4 is converging on a point, so the system is moving towards a stable equilibrium.

(a)   **(b)**

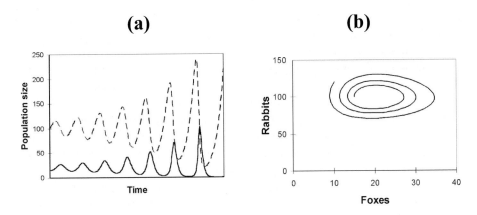

Figure 5-4. Population changes in a model of foxes preying on rabbits. (a) A plot of population size for foxes (solid line) and rabbits (dashed line) through time. (b) The trajectory of the system through time. In this case the system is unstable, so the trace spirals outwards until the fox population crashes to zero (not shown).

It is important to realise that dynamic systems are not confined to systems that take numerical values, such as logistic growth or the foxes and rabbit model. Any system that passes from one state to another is a dynamic system. That includes all computers as well as (say) animal behaviour or the working of the brain. So all of these systems potentially exhibit properties similar to those we are discussing here.

5.2.1   Equilibrium and stability

Equilibrium is a state of balance. A system in equilibrium shows no tendency to change of its own accord. A variable is said to be an *equilibrium* point if there is no tendency for its value to change through time. That is, the forces that act to change it are balanced out.

Place an orange at the bottom of a fruit bowl and it will stay there. It is in equilibrium. If you roll it up the side of the bowl and release it, then it will roll back down to the bottom again. So the equilibrium is stable. In general, any equilibrium point is said to be *stable* if the system tends to return to the equilibrium when displaced from it (Figure 5-5). This condition means that the rate of displacement away from the equilibrium point must be opposite in sign from the direction of the displacement. If the system continues to move away from an equilibrium point when a displacement occurs, it is called *unstable*. If a displacement leads to no further movement, it is termed *neutral*.

One of the simplest ways to visualise the above ideas is to imagine the system as a ball rolling across a landscape (Figure 5-5). When the ball stops rolling, we say it is in equilibrium: it has no tendency to move of its own accord.

Suppose the ball has rolled down into a hollow in the grass. Outside forces, such as the wind, might push the ball up one side of the hollow a little, but it quickly rolls back down into the deepest part of the hollow. In this case, the ball has found a stable equilibrium. The walls of the bowl provide negative feedback.

Note that for most systems, a very strong perturbation can push the system away from an equilibrium point, just as a really strong gust of wind might push the ball out of the hollow and back into the wider landscape. There are limits to stability. That is to say, stable equilibria are usually relatively, but not absolutely stable. Now suppose that, instead of falling into a hollow, the ball has been placed at the summit of a steep hill. Even a slight nudge will tend to send it rolling down the hill and far away. Sitting poised at the top of the hill, the ball is at an unstable equilibrium point. The sides of the hill provide positive feedback.

Lastly, imagine not a ball, but a heavy cube sitting on a flat plane. It will move in response to an outside force, but halt as soon as the force is removed. It is in neutral equilibrium. Neutral equilibrium is not a likely scenario in any ecological context; it is included here merely for completeness.

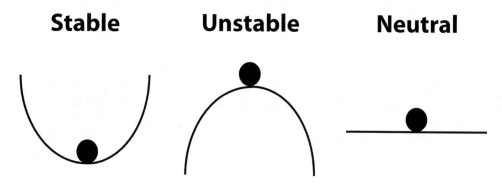

Figure 5-5. Three kinds of equilibrium. In each case a ball starts at rest on a surface, as shown, and does not move unless pushed. Left: when moved, the ball rolls back down the slope to its initial position. Centre: if moved, then the ball continues to roll away down the slope. Right: if moved, the ball will remain in its new position.

Formally analysing the stability of an ecological system can lead to useful insights. In many parts of the world, dryland ecosystems are turning into deserts. Because drylands include a high proportion of farms, this loss is devastating for farming communities and contributes to famines and poverty in the developing world. For many years, the reason for this widespread pattern remained unclear.

In 1993 geographer Jonathon Phillips analysed the stability of a well-known systems model of dryland ecosystems (Phillips 1993). He was able to show that drylands are likely to be inherently unstable because of positive feedback between key factors such as vegetation cover, light absorption, rainfall, soil moisture and erosion. Perturbations to dryland ecosystems, Phillips showed, would not average out, but would persist and grow. The dryland is like a ball poised at the top of a hill: even a small disturbance can magnify to push it far from its original state. Thus, chance disturbances can easily cause a dryland to become desert, but will rarely if ever cause a desert to revert to dryland. Sustainable management of drylands therefore requires close monitoring and quick correction of small disturbances.

Stability analysis has also been used to understand the dynamics of food webs and ecological communities. We will look at these systems in Chapter 7.

5.2.2   Transients and attractors

As we saw above, a dynamic system is stable if its behaviour converges over time to a fixed point. However, not all systems behave in such a simple way. In many systems the behaviour converges to a cycle, instead of a single equilibrium point. Cycles of this kind are called *limit cycles*.

Equilibria and limit cycles are both cases of a class of behaviour known as *attractors*. In general, an attractor is any set of states to which a system is "attracted". In an equilibrium that set consists of a single point. In a limit cycle the system passes through the set in sequence, then repeats. Later we shall see that chaotic systems also have attractors, but with more complicated structure.

The term attractor is used because the system tends to be attracted towards those states. In like fashion, there are sometimes states - *repellers* - that repel the system. A characteristic of an attractor is that once a system falls into an attractor, it stays there unless subject to some external disturbance. Thus stable equilibria are attractors and unstable equilibria are repellers. In the desert-dryland system mentioned above, deserts are attractors, while drylands are repellers.

Starting from an arbitrary initial state, a dynamic system will normally pass through a series of transient states until it falls into an attractor (Figure 5-6). A system may contain more than one attractor. If so, which attractor it ends up in depends on where it starts. In general, any attractor is surrounded by a region known as its *basin of attraction*.

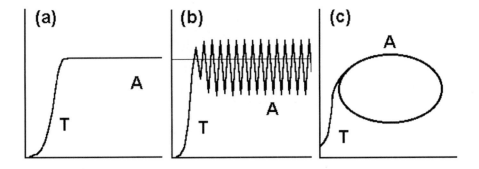

Figure 5-6. How dynamic systems work. In each case the label T denotes transient behaviour and the label A indicates an attractor. (a) An equilibrium is a point attractor. (b) Limit cycles plotted against time. (c) Limit cycles in the foxes and rabbits system (plotted in state space as foxes versus rabbits).

We can imagine a dynamic system as a ball dropped into a rugged landscape and pushed at random by the wind. The basin of attraction behaves like the slope of a valley: once the ball has passed the rim, it tends to roll downwards, and becomes increasingly unlikely to escape the attractor as it falls. Which valley the ball finally lands in may be impossible to predict, because it depends on its precise initial position, as well as on the magnitude and direction of random perturbation by the wind.

5.2.3   *Sensitivity to initial conditions*

So far we have discussed dynamic systems that are either in equilibrium or have fallen into a limit cycle. A common feature of non-linear systems is that when they are far from equilibrium, their behaviour often diverges, rather than converging. This property, which makes them sensitive to small changes in their initial conditions, was discovered independently by several researchers.

Edward O. Lorenz was trying to simulate weather systems on an early computer. He was puzzled to see that when he restarted a simulation, using results he had printed previously, the output rapidly diverged from his earlier findings. After just a handful of iterations, the model behaved in a completely different way. At first, Lorenz thought he must have accidentally entered an incorrect digit; but he had not. Intrigued, Lorenz repeated his experiment – and once again got completely different results (Figure 5-7). There could be only one explanation. The difference was caused by tiny changes introduced when numbers in the simulation were rounded off for printing.

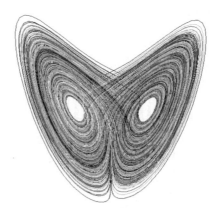

Figure 5-7. The Lorenz attractor. This (three-dimensional) figure shows the trajectory followed by Lorenz's weather model. Within the attractor, the position is unpredictable, but if a system starts outside, it soon falls into the attractor. The attractor itself is densely braided. If you restart the system at any two points close together, then they soon drift apart.

Around the same time, mathematical biologist Robert May was experimenting with discrete logistic models of rapid population growth. Varying the initial population size by just a tiny fraction, May found that in models that were completely identical apart from small differences in initial conditions, the magnitude and timing of fluctuations in population size rapidly diverged. Even the most miniscule variation would effectively make population size impossible to predict within a few tens of generations (Figure 5-8).

The sensitivity of non-equilibrium systems to changes in their initial conditions has many implications. Perhaps the most important is that the detailed behaviour of such systems is unpredictable. For example, we can make reasonably accurate predictions of the weather up to a couple of days ahead. But beyond that limit, real weather patterns become increasingly different from our predictions. Similarly, if we have an excellent understanding of the ecology of a species and it lives in a predictable environment, we may be able to make reasonably accurate predictions of population size across a generation or two, but as time progresses, our confidence rapidly declines and predictions become meaningless. Economic projections are likewise limited.

The essential reason for this unpredictability is that we can never measure every aspect of weather or populations or economic activity with total accuracy. The sensitivity of the system means that these errors rapidly become magnified. Likewise, small random fluctuations add tiny errors that are also magnified by the system. This unpredictable behaviour is the domain of chaos theory.

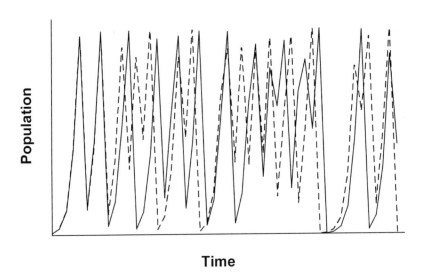

Figure 5-8. Sensitivity to initial conditions in discrete logistic growth. The two populations are identical, except that the population represented by the dashed curve started with one extra individual. Notice that after just a few cycles the dotted line deviates from the solid line.

5.2.4   The onset of chaos

Chaos theory deals with systems that are deterministic, but whose long-term behaviour is unpredictable nevertheless. As we saw in the previous section, this unpredictability is a result of the sensitivity to initial conditions. By deterministic, we mean that no outside perturbation is necessary to generate unpredictable behaviour: it emerges from our finite knowledge of initial conditions.

Chaotic behaviour may be unpredictable, but it is constrained. To understand how it is constrained, first consider constraints in ordered conditions. As mentioned above, ordered systems contain either equilibrium points or limit cycles. Both of these may be considered as attractors, in the sense that if the system starts in some arbitrary state, or else is disturbed, then it is attracted steadily closer to the equilibrium or to the limit cycle.

The equivalent of an equilibrium point or a limit cycle for a chaotic system is a *strange attractor*. The difference is that a strange attractor is a far more complex structure (Figure 5-8). Whereas a limit cycle is a closed loop, a strange attractor never crosses itself. That is, the system is never in exactly the same state twice. It may exhibit quasi-cyclic behaviour at times, but the cycles are never closed and separate strands of the strange attractor may be braided together with infinite density.

The extreme sensitivity of chaotic systems to initial conditions is a consequence of this braiding: changing the initial state even slightly moves a chaotic system onto a different braid. It is sometimes called the *butterfly effect*: in theory, even the minute air disturbance caused by a butterfly beating its wings could tip large-scale weather patterns in unpredictable ways.

When first discovered, chaos seemed to be a mysterious and puzzling anomaly in otherwise orderly systems. However, it is now known that the transition of a dynamic system from equilibrium to chaos is itself a very well-defined process.

Typically, as we change the value of a particular number, called the *order parameter L*, the onset of chaos is associated with a transition of behaviour from stable equilibria, through limit cycles of increasing period, to chaotic variation associated with a strange attractor.

This transition occurs via a process known as *period doubling*. That is, at certain critical values of L, the period of the limit cycle doubles (Figure 5-9). Furthermore, after the period starts doubling, the increase in L required to produce the next doubling steadily decreases[5].

We can represent the transition to chaos using a bifurcation diagram (Figure 5-10). This diagram plots the attractors of the system against L. In the example shown the system settles into equilibrium when L is small. This value changes as L increases, so it appears as a single line in the diagram (Figure 5-10). However as L increases the system switches to limit cycles of period 2. So the line bifurcates into two lines - one for each point in the limit cycle. As L increases again, the

---

[5] In the case of discrete logistic growth $x'=Lx(1-x)$, for example, the period is 1 (i.e. an equilibrium point) for $2 < L < 3$, but in the interval $3 < L < 4$, the period doubles to produce limit cycles of period 2, 4, 8, 16, ... until the period becomes effectively infinite at the point $L \sim= 3.93$ (Figure 5-10).

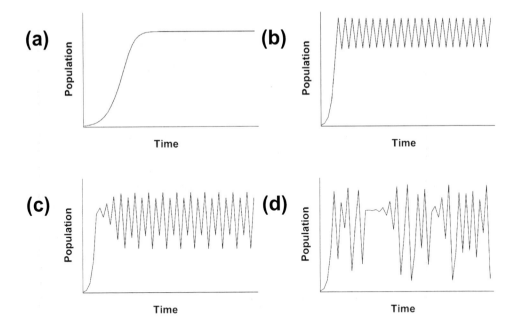

Figure 5-9. The onset of chaos in a model of logistic growth. (a) When the growth rate is low, the population rises smoothly to an equilibrium level. (b) At higher rates the system settles into cycles of period 2. (c) As the growth rate increases still further the cycling period doubles. (d) Eventually this period doubling leads to chaotic behaviour.

system bifurcates again and again, so that 2 lines become 4, then 8 etc (Figure 5-10).

Notice that at any bifurcation (Figure 5-10), the pattern from that point onwards resembles the overall pattern, except that it is squashed. This resemblance is no coincidence: if the two axes are rescaled by appropriate amounts, then the pattern is identical to the overall pattern. This decrease occurs in a fixed way with the changes in L between successive sets of doublings falling in the fixed ratio[6] of 4.66920.

Although the above theory suggests that chaotic behaviour should be a widespread property of ecological systems, demonstrating chaos in practice has proven difficult. One reason for this difficulty is that it is difficult to control populations sufficiently to rule out the influence of external factors causing their unpredictable dynamics. There has been no simple way to test whether a natural system is chaotic or merely random   (Poon and Barahona 2001). Another possibility is that chaos may tend to be damped by natural selection, as populations with chaotic dynamics are vulnerable to extinction because of their

---

[6] This is known as Feigenbaum's ratio after its discoverer.

-    *resistance*—the degree to which a variable is changed, following a disturbance;
-    *variability*—the variance of the variables over time.

To understand the stability of entire ecosystems, we need to look more closely at species interactions and the effects of landscapes. We will look at these and other issues that influence the stability of whole ecosystems more closely in Chapter 7.

5.3.3 Does a balance really exist?

During the latter part of the 20th century, the assumption of equilibrium in ecology was increasingly called into question. Elton (1930) was one of the first to question this assumption:

> *"The balance of nature does not exist, and, perhaps, never has existed. The numbers of wild animals are constantly varying to a greater or lesser extent, and the variations are usually irregular in period and always irregular in amplitude."* (Elton 1930).

This sceptical attitude was been reinforced by the realization that many kinds of ecosystems are subject to chronic disturbance, such as drought or fire (for example, Bormann and Likens 1979; Noble and Slatyer 1981). Some of these ecosystems actually depend on disturbance to persist.

Theories that rely on the assumption of equilibrium have also been called into question. For instance, since Macarthur and Wilson proposed the theory of island biogeography, several authors have highlighted its short-comings (for example, Brown and Lomolino 2000, Lomolino 2000, Whittaker 2000). The number of species found on a particular 'island' is not limited by area alone. It is more likely that many factors combine to determine the number of species that can coexist within a particular patch of landscape. These factors include the variety of habitats and the availability of food, light and water, as well as the ecological character of the species concerned. In some cases, variations in these factors across a landscape do correspond with area, but often they do not.[8]

In the next chapter, we will see that pollen profiles, which record the history of vegetation change over long time periods, provide a different view of the stability of ecosystems.

---

[8] The influence of island area and isolation, species diversity is also known to:
-    exist in a non-equilibrium state (Whittaker 1998);
-    be simultaneously influenced by phylogenetic diversification (phylogenesis) (Heaney 2000); and
-    be influenced by the different 'island' characteristics other than just area and isolation, for example habitat diversity, level of disturbance (Fox and Fox 2000).

CHAPTER 6

POPULATIONS IN LANDSCAPES

Mangroves growing along an estuary, northern Australia. In this chapter we show how a trade-off between salt tolerance and growth rate can create bands of different species on salinity gradients.

6.1 ONE POPULATION OR MANY?

Halfway between Sweden and Finland, and separating the Baltic Sea from the Gulf of Bothnia, lies a vast group of islands known as the Åland archipelago. Consisting of more than 6,500 islands, Åland is home to a species of butterfly known as Granville Fritillary (*Melitaea cinxia*), which inhabits the small patches of dry meadows that are found on many of the islands[1]. For the most part, the butterflies on each island form a distinct population. About 60-80% of the butterflies spend their entire lives on the island where they are born. So the butterflies living on each island are almost isolated from the rest of the species.

However, the population on any single island is never large enough to persist on its own. There is always a high risk that it will become extinct, through one means or another. Fortunately, however, the many island populations are not completely isolated from one another. A small percentage of butterflies stray far from home. When the butterflies become extinct on any island, butterflies from other islands soon appear and re-establish the species locally. Mass extinction of the butterflies from all islands is highly unlikely, because the subpopulation on each island is sufficiently isolated to be unaffected by any disaster that might kill off butterflies on another island.

The history of the butterflies of the Åland archipelago shows that the idea of a population is not quite so simple as it may first appear. The first problem we face in trying to understand a real-life population is identifying where it begins and ends. What are its boundaries within the landscape? If you find a frog living in a swamp, and frogs of the same species living in another swamp nearby, are you seeing one population or two? It is rare to find a group of animals or plants living within an easily defined area except, perhaps, on islands in the middle of an ocean. Most habitats and resources are spread unevenly, as also is the intensity of competition, so most populations occur in patches of high and low density across a landscape. But these patches are usually not totally isolated from one another (Figure 6-1): the different *subpopulations* interact to some extent (Simberloff et al. 1992).

This picture of populations being made up of subpopulations leads to the idea of metapopulations. A *metapopulation* is a 'population of populations'. It is a mistake to focus on a single subpopulation without considering how it is influenced by other subpopulations in the landscape. The Åland butterflies provide a classic example of a *metapopulation*[2].

The theory of metapopulations deals with the colonisation and extinction of individual species within archipelagos and patchy environments. The idea is closely linked with processes such as population turnover, extinction and establishment of new populations (Hanski and Gilpin 1991). Small populations are

---

[1] *Melitaea cinxia* distribution on the Åland islands as an example of a metapopulation. Much later, studies by Hanski (1997), and by Moilanen and Hanski (1998), investigated their population dynamics in more detail, and used this species to demonstrate that modelling metapopulation dynamics can help elucidate the effects of habitat patch size, degree of isolation and habitat quality.

[2] Metapopulation theory was originally developed by Richard Levins (1969).

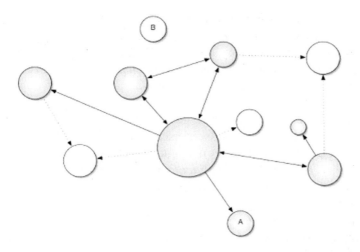

Figure 6-1. A metapopulation in a landscape. The sizes of the circles indicate the level of opportunity that a given patch of landscape offers individuals (resources balanced against, for example, level of competition). So circle size would normally reflect actual or potential size of a subpopulation. Unshaded patches have no individuals present at this point in time, but could have them in the future (and may have had them in the past). Arrows represent the direction in which genes flow between subpopulations (via immigration, pollination, and so on). Some subpopulations are sources of genes for the metapopulation ('source populations'); others are 'sink populations' (for example, subpopulation A). The level of connectivity within the metapopulation is represented by solid lines (for current levels) and broken lines (potential connectivity). In this example, population B does not interact with other subpopulations and is therefore not part of this metapopulation.

sensitive to extinction, and local extinctions are regular events. Survival of a local population depends on the interaction with surrounding populations.

What is important here is how well connected the different patches or islands are. If no individuals move between different patches, then you have a collection of isolated populations. But if isolation is not complete, if the patches do interchange individuals, and if the movement between populations is sufficient to recolonise a patch after the species becomes locally extinct, then, like the butterflies of Åland, regional extinction of the entire group of populations (a metapopulation) becomes unlikely.

The theory of metapopulations is particularly useful for understanding processes that take place within fragmented habitats. Because the idea of metapopulations requires a landscape approach, it reflects the processes that occur within ecosystems better than a simple population model. The degree of connectivity within a population is intimately related to the rates at which immigration and emigration take place between patches. These rates can directly determine extinction rates, and establishment rates, of the subpopulations (Hanski and Gilpin 1991).

Metapopulation theory is useful to landscape managers. They need to know how connected a population needs to be to maintain a genetically-viable population. Metapopulation theory also justifies the protection of areas within a landscape that offer potential habitat for a target species, even though no individuals may be present at the time. Given that the process of fragmentation is a change in the spatial arrangement of a particular habitat, it is not surprising that most ecologists have readily adopted metapopulation models based on spatial distribution of populations on a landscape scale. Indeed, the degree of interaction among the subpopulations will affect the dynamics of the entire population.

Leadbeater's Possum (*Gymnobelideus leadbeateri*) is an endangered tree-living mammal in the Central Highlands of Victoria, Australia. It is native to the montane ash forests of the region. Until re-discovered in 1961, it was thought to be extinct. The trees that provide suitable nesting hollows (200 or more years old) are currently in rapid decline. Devastating fires in 1939 and subsequent salvage logging has resulted in a significant period of 140 years before the 1939 regrowth can provide suitable nesting sites for this species. The decline in the population caused by this bottleneck is estimated at 90%. A metapopulation model was used to estimate the probability of extinction of this species (Lindenmayer and Possingham 1995, 1996). Single isolated patches of less than 20 hectares were predicted to be highly susceptible to extinction, with survivorship increasing with patch size. Furthermore, the importance of several 50-100 hectare forest blocks for the survival of the species became most obvious when simulating the possibility of catastrophic events wiping out populations.

From the above examples, it should now be clear that the interactions that take place within a population are more complex than most people would at first think. In this chapter, we will attempt to understand the complexity of individual populations within a landscape, by considering population growth and the distributions of species, and examining how these factors are directly affected by competition and other interactions. Understanding the complexity of interactions among populations (i.e. different species), requires careful study of the dynamics of communities and ecosystems, which we will look at in more detail in later chapters.

6.2 SPATIAL DISTRIBUTIONS

The ways in which organisms are distributed in a landscape can be very complex. As we saw in Chapter 2, the sheer number of interactions among different species living in the same area can be huge. The physical environment may also vary. However, spatial processes and interactions also have a big influence on distribution patterns.

Plants and animals in a landscape compete for space. In a long-untouched environment, new arrivals are often elbowed out by older and larger residents. Seedlings falter in the shade of forest giants. Young animals wander, driven from every habitable nook by dominant adults. But all this changes when we take mortality into account. Wildfires, storms, floods and other disasters occasionally strike, ripping away established inhabitants and leaving patches of naked earth ripe for colonisation. Even the death of a single large tree creates an opening in the forest canopy, an opportunity for dormant seedlings to race towards the light. Meanwhile, decaying tissues offer temporary homes for a whole suite of specialist decomposers that will themselves leave behind fertile soil for the community's regeneration.

The colonists will not come at random. Rather, most will be descendants of neighbouring plants and animals. A few will come from longer distances, travelling along linked areas of habitat suitable for their kind, or perhaps drifting with wind or currents. All will come from sites that are in some sense connected with the new site. It is for this reason that understanding the pattern of connections between sites in the landscape is essential for predicting long-term changes in the composition of the community.

Processes that involve movement across a landscape are responsible for many kinds of patterns in the distribution of vegetation. Figure 6-2 gives some examples of the patterns that can result. In these simulation experiments, we set up a very simple cellular automaton (CA; see Chapter 3, Section 3.1.2) landscape model in which each site (cell of the CA) is occupied by either of two types of plant. The cells representing these plants are drawn either in black or white on the maps. Each cycle of the model represents a breeding season. For the sake of simplicity, we make several assumptions. For instance, we assume that the two species of plants are identical in all their vital attributes: they mature the same way, they grow at the same rate, and so on. They are identical in every way, except that they just happen to be different species. We assume that the plants are relatively short-lived, so free sites are appearing continually. We also assume that the plants become mature (that is, they start producing seeds) almost immediately and that there is no shortage of pollen and pollinators. The advantage of keeping the experiment simple is that it helps us to see the wood rather than the trees, so to speak. It focuses on one or two particular mechanisms without the results being confused by the interaction of many different effects.

In the first experiment (shown in Figure 6-2a), we assume that seeds from the plants can spread anywhere in the entire area. The result is a random distribution in which plants are scattered evenly across the entire landscape. In the second experiment, we assume instead that seeds can only spread a short distance from the parent plants. This model is consistent with, say, seeds falling around the parent tree. The resulting distribution of plants (Figure 6-2b) is very different from the previous one. Both of the 'species' form tightly clumped distributions.

If we introduce disturbances, such as fires, into the above models, the resulting plant distributions are different again (Figure 6-2c). Fires burn out entire patches of forest, which are then quickly invaded by plants from the surrounding area. The nett effect is to make the clumps that we saw above coalesce into large patches. In other words, the landscape becomes dominated by large areas of uniform composition. Again, we have made some simplifying assumptions. In particular, we ignore any seeds that might be stored in the soil and germinate after clearing.

In the final experiment, we look at what happens when the above effects (local dispersal plus clearing by fire) combine with environmental patterns. Suppose that an environmental gradient runs across the landscape. Gradients of this kind are

extremely common. It might be a decrease in soil moisture as you move uphill
from a river. It might be an increase in salinity, from fresh water to brackish water,
across a mangrove swamp at the mouth of an estuary. It might be a decrease in
temperature as you move up the side of a mountain.

To see any effect, we must now introduce a difference between our two
competing plant populations. We suppose that they prefer opposite ends of the
environmental spectrum that we have introduced to the landscape. With this one
difference, and everything else as in the previous experiment, the resulting
distributions change yet again (Figure 6-2d). The patches that we saw form in the
previous experiment now become fixed in space at either end of the gradient. They
form two distinct regions. Notice also that the two regions do not blur gradually
from one species to another. Instead there is a sharp boundary. If you walked
across the landscape, you would first see one species totally dominant, and then
suddenly it would disappear and the other species would take over.

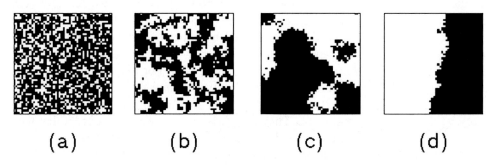

(a)   **(b)**   **(c)**   **(d)**

Figure 6-2. Spatial patterns in the distribution of plant species induced by dispersal.
(a) Seeds can disperse anywhere; (b) seeds spread from local sources; (c) when fire
is added to the system the clumps in (b) coalesce to form patches; (d) the effect of
environmental gradients is to fix patches in space. The results were derived from
cellular automata simulations of forest ecosystems. The initial distributions were
random in each case (after Green 1994b).

6.3 PATCHES, EDGES AND ZONES

The traditional approach of ecologists to landscapes has been to treat them as the
settings for environmental variation. Changes in ecosystem structure and
composition are seen to reflect changes in environmental conditions from place to
place. This approach is embedded in standard field methodology. Sampling
methods traditionally employ the quadrat (usually a square of fixed area) as the
unit of sampling[3].

A well-developed body of statistical theory for experimental design and
interpretation has grown up around quadrat sampling (for example, Pielou 1974).
In particular, ecologists study spatial variation by examining differences between
sets of quadrats. Randomly arranged sets of quadrats are the standard way to
sample uniform environments. A sequence of quadrats (transects) provides a way

---

[3] See Chapter 3.

to study variations along environmental gradients such as changes in soil moisture on a hillside or increasing salinity in an intertidal swamp.

One of the most important results to spring from transect studies is that environmental variations alone do not suffice to explain the spatial distributions of organisms. Research by quantitative ecologists showed that landscape interactions, such as competition and dispersal, play important roles too (for example, Pielou 1974). For instance, competition between species often leads to truncated distributions along an environmental gradient. The significance of these findings is that ecosystems are not controlled in simple fashion by external (i.e. abiotic) factors. Internal, biotic interactions within a system play an important role too. These biotic interactions are likely to be non-linear and complex. In the sections that follow, we will look at how research into the interactions between species has led to a new understanding of their geographic distribution for two ecosystems.

Salt of the earth

In estuaries, fresh water from rivers mixes with salt water from the sea. At the river mouth, the water is almost as salty as the sea. Moving further up the estuary, the amount of salt gradually declines, until eventually the water is fresh. So the water in an estuary provides a good example of an environmental gradient in space, with salinity being the variable concerned.

If you go walking through a mangrove swamp, you might notice a curious thing. As you progress towards the shoreline, you pass through successive bands of mangrove that are subtly different. In fact, each band of mangrove is a different species. Observing these zones, ecologists naturally assumed that each mangrove species grew in the area where the salinity was ideal for it. As in other cases, inter-species competition would explain sharp transitions at the zone boundaries.

As it turned out, this obvious and intuitive theory was almost completely wrong. Research by ecologist Marilyn Ball produced a surprising result. Instead of different salinity levels being optimal for different species, it turned out that the optimal salinity level for *every* mangrove species was 10% sea water. What differed was the tolerance of the species to extremes of salinity. Some could grow only in very low concentrations of sea water. Others could survive much greater levels of salinity.

But there is a price, a genetic cost, for tolerating high salinity levels. The price is that these species have to devote a lot of physiological effort towards dealing with salt. The trade off is that they grow slowly. And the greater the salt tolerance, the slower the trees grow. Slow growth rates mean that other, faster growing species can beat them to take over spaces in the swamp. So the only places where slow growing species can dominate are at locations where faster growing plants cannot grow. That is, they are confined to the extreme high end of the range of salinities they can tolerate. The point is that any one of the species could easily cover all of the areas with low salinity, but competition restricts them to a small zone.

They prepared these samples for pollen analysis by a series of chemical and physical treatments that removed all unwanted components from the samples. The most dangerous step in the processing was to remove silica (sand) by heating the sediment in hydrofluoric acid, one of the most corrosive substances known to science. It actually dissolves glass, and flesh! Another step involved boiling samples in an explosive mixture of concentrated sulphuric acid and acetic anhydride.

From the concentrated pollen "soup" David placed small drops on glass slides. Viewing these samples under a powerful microscope, he laboriously counted pollen grain after pollen grain, to find out exactly how much of each kind of pollen was present in the sample. Counting pollen is a slow, exacting procedure. Over the following twelve months the forest history around Everitt Lake gradually emerged, sample by sample, time slice by time slice.

When finally complete, the pollen record showed that over a 12,000 years period, the ecosystem around Everitt had undergone dramatic changes, from tundra in early postglacial times, to coniferous forest (Figure 6-4), deciduous hardwoods, to mixed forest, and finally to the appearance of farming and agriculture in recent times. The forest history at Everitt Lake was fairly typical of sites in the region. However, because the sediment was sampled and counted in finer detail than any previous pollen cores from the region, it provided new and

Figure 6-4. Spruce-larch conifer forest in southern Nova Scotia, Canada.

revealing details about the processes involved in forest change.

Palynology, as the systematic study of preserved pollen is called, began around 1900. The Swedish geologist Lennart von Post, often called the "father of Quaternary pollen analysis," had the crucial idea that changes in the quantities of preserved pollen should reflect changes in the sizes of the corresponding plant populations. As we saw earlier, what palynologists traditionally do is to collect mud from a lake or bog by pushing a long tube into the bottom sediment.

The palynologist extracts samples from progressively deeper, and therefore older, sediment levels. Processing these samples reveals the changing proportions of different kinds of preserved pollen. When plotted in sequence (a pollen diagram) the results show how a plant community changed through time (see Figure 6-5).

One of the first and most important findings of palynology was that vegetation histories, as recorded by pollen diagrams, are remarkably consistent, even across vast regions. This meant that the preserved pollen was telling a true story about environmental history, and not just local sediment changes. At the same time, the discovery also raised a new set of questions. What could have led to forest changes being so similar at so many different places? The obvious candidate was climate change. Early pollen studies were all carried out in regions that had been covered by ice during the last ice age. So as the ice melted and disappeared, each site was subjected both to climate change, and to species migrations.

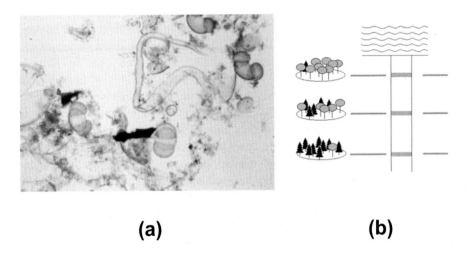

(a)   **(b)**

Figure 6-5. (a) A typical view from a pollen slide, showing pollen grains, miscellaneous plant matter and specks of charcoal (the black fragments). (b) A cross section of sediment from a lake bottom, showing the varieties of past forests that might be represented. The deeper the sediment is in the column, the older the forest.

Subsequent research has produced many ecological insights about past vegetation change, including the development of agriculture, environmental degradation by early civilisations, and the role of fire in long-term forest change (Green 1982, 1987; Tsukada 1982; Davis 1976; Webb 1979).

6.6 GALLOPING TREES?

About 18,000 years ago, most of eastern North America was covered by huge ice sheets, over a kilometre thick in places. As the ice melted, vast areas of land again became ice free. Plants migrated into these regions and forests appeared.

Regional comparisons of forest histories derived from preserved pollen data have made it possible to unravel the patterns of postglacial plant migrations. In North America, palynologists such as Margaret Davis and Tom Webb compiled data from hundreds of individual sites (Davis 1976; Webb 1979). Using this data, they reconstructed the northward migration patterns of forest populations as they followed retreating ice sheets. These maps portrayed the northward migrations of tree species following the last ice age as a steady, continuous flow.

For the most part, people have accepted the maps arising from these migration studies at face value. That is, everyone has simply assumed that the contours showing arrival times of different plants at different locations should be taken literally (for example, Davis 1976; Webb 1979; Bennett 1983, 1985; Delcourt and Delcourt 1987; Williamson 1996). The truth was not quite so simple. The entire process was more complex, with the interactions between different competing populations having an enormous influence on the course of events.

The story begins with a mystery. A surprising feature of tree migration patterns is that many populations advanced across the landscape much faster than anyone had thought biologically possible. Oak trees, for instance, produce large acorns, which normally fall within a few metres of the adult tree. Yet in Britain, populations of oak trees were able to spread northwards at rates of hundreds of metres per year. Overall speeds achieved by migrating trees were typically around 200 metres per year, but were sometimes as high as two kilometres per year. It was almost as if the trees had sprouted feet and walked. And in a sense, perhaps they had. Such speeds may be startling if we restrict our thinking to seeds falling from a tree and settling on the ground nearby, but animals and birds sometimes carry seeds great distances. Storms may be even more important in long distance transport. At first sight, it may seem impossible that wind could pick up a large seed such as an acorn and move it anywhere. However, hurricanes at the right time of the year can easily rip entire branches off trees and carry them kilometres in one giant step.

Studies of very detailed pollen records in eastern Canada have revealed that the migration maps are misleading (Green 1982, 1987). It turns out that populations do not move forward in broad fronts at all. Instead, storms and other means of transport enable small patches of trees to gain a foothold well outside their normal range. In Nova Scotia, for instance, traces of virtually all the tree species that make up today's forests appeared very early in the postglacial forest

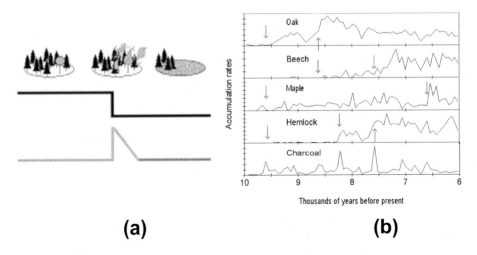

(a)   **(b)**

Figure 6-6. Cataclysmic change in postglacial forests (adapted from Green 1987). (a) A time slice showing changes in forest composition that might accompany a fire and the way this event might appear in the preserved records for pollen (black line) and charcoal (grey line). (b) Pollen and charcoal records (those presented here are from Everitt Lake, Nova Scotia) show that competition from established species suppresses invaders. The fundamental assumption of pollen analysis is that changes in the rate at which pollen accumulates in sediment records indicate corresponding changes in the plant population that produced the pollen. Major fires (indicated by sharp peaks in the charcoal record) clear large tracts of land, remove competitors and trigger explosive growth (arrows) of invading tree populations. Especially clear for oak, beech and hemlock is the role of fire in promoting invasions of new species.

record (Figure 6-6). But what happened then was that the populations that were best adapted to the prevailing conditions out-competed other species, preventing them from expanding. In some cases, seed populations, such as beech and hemlock, were suppressed for thousands of years, even though they were perfectly able to cope with the prevailing climatic conditions.

Eventually, each suppressed population underwent rapid expansion (Figure 6-6). Disturbances played an important part in this process (Green 1982). Small, regular fires helped to maintain some community types by excluding competing species, whereas large fires, under changing climatic conditions, triggered sudden changes in plant communities, especially invasions, by reducing the competitive advantage of established populations. In each case of major vegetation change, widespread fires triggered the event. In other words, the migration maps mentioned earlier were not charting species arrivals at all. Rather, what they trace are the explosions of populations from isolated seed trees that were already present in each area. So, in the debate over transport rates, researchers have really focused on the wrong question. The real question is not how trees manage to migrate so rapidly, but why they migrate so slowly!

Further pollen studies have since confirmed that this model of tree migration correctly represents the processes that took place. For instance, Keith Bennett (1985) and Matsuo Tsukada (1982) have shown that models of exponential population growth fit the postglacial expansions of many tree populations in Britain, Japan and North America. Also, later pollen studies have shown that fires have triggered abrupt changes in forest composition in many other parts of the world[6].

---

[6] For example, Chen (1988) found similar events in the forest history of north Queensland.

of such interest. The polar regions, mountain tops, and ocean depths are all revealing. But we do not need to go to the ends of the earth to find extremes. For living things, the important extremes are the environmental factors that limit their distribution. We see such limits everywhere in the form of environmental gradients. Environmental gradients may occur either in space or in time. Spatial gradients include such environmental features as altitude, soil moisture, salinity, and nutrient levels. Environmental gradients in time include such well-known phenomena as climatic change (for example, increasing temperature or rainfall), increasing aridity, and encroaching salinity.

Environmental gradients hold great interest for ecologists because they create a rich variety of landscape patterns and conditions that favour different species (see Chapter 6). In numerous studies during the 1960s and 1970s, the Canadian ecologist Chris Pielou showed that environmental gradients expose the mechanisms of inter-species competition to view (Pielou 1969).

One of the most telling effects of environmental gradients is to restrict the range of species (Figure 7-3). If there were no interactions at all between species that have overlapping environmental responses, then there would be a smooth transition between their distributions. The range of one species would merge gradually into the range of the next (Figure 7-3a). However, species do interact. Among other things, they compete for space. Within a local area, the population with greater numbers has a marked advantage when competing for vacant growing sites. They therefore tend to exclude competing species. The result is that on a gradient, species distributions tend to become truncated (Figure 7-3b), with sharp boundaries marking the edge of their territory. In Chapter 6, we saw that spatial processes play a large part in creating these truncated distributions. Simulation experiments (Figure 6-2d) confirm that a combination of patchy disturbances (for example, fires) and short-range seed dispersal are capable of creating sharp

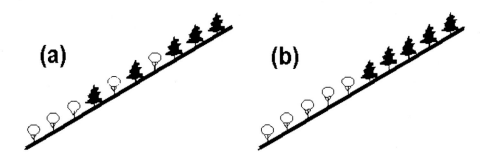

Figure 7-3. Competition on an environmental gradient. (a) In the absence of competition, the range of one species would merge into the range of another. (b) The effect of competition is to create abrupt transitions (zones) between the range of one species and the range of its competitors.

boundaries between ranges when pairs of species compete on an environmental gradient. The same effect can happen when environmental gradients occur in time rather than in space. The most common examples are long-term shifts in climatic variations in, for example, the annual or seasonal changes in rainfall, or the average temperatures.

To learn what is likely to happen during competition over time, we can carry out a simulation experiment in which we replace an environmental gradient in space with a gradient over time. A simple experiment consists of making this replacement with the same kind of model that produced Figure 6-2d. That is, we have two populations competing with one another. Initially, one is greatly favoured by the prevailing environment and outnumbers its competitor by an order of magnitude. In all other respects, we suppose that the two species are identical. Over time, the environment changes until ultimately the roles are reversed. If dispersal is not limited, then the change in population size is smooth and gradual (Figure 7-4a). Notice, however, that the population that is initially greater in size remains so for a long time after the environment favours its competitor.

In contrast, if dispersal is limited for both species (for example, by plants requiring local seed sources), then the changeover in composition takes a very different course. In this case, disturbances trigger massive, sudden changes in the sizes of both populations (Figure 7-4b). This is exactly the kind of population change that has been observed in pollen records (for example, Figure 6-5b), and which has been attributed to a combination of disturbance and changing climate (Green 1982). What the simulation models do is to suggest that spatial distributions, as well as dispersal, play a crucial role as well.

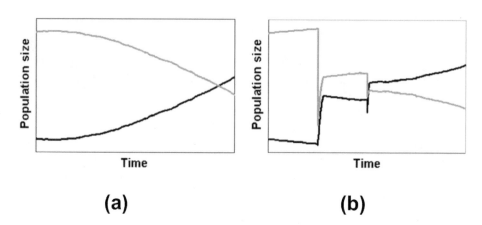

(a)   **(b)**

Figure 7-4. Competition on an environmental gradient in time. In this model scenario, two species compete for territory. One species initially dominates the region, and suppresses its competitor. (a) 'Global' dispersal. (b) 'Local' dispersal.

7.3. WHAT IS AN ECOLOGICAL COMMUNITY?

The word 'community' can mean different things to different people. Even ecologists use it in different ways. For instance, ecologists sometimes refer to groups of 'similar' species in an area as a community, such as the 'bird community' of a forest. However, when ecologists speak of a community, they normally mean a group of populations that live in the same area and interact in some way, such as through competition, predation, and so on.

So communities consist of populations that interact with one another. That is, the populations form a complex system—an ecosystem. But is it enough that plants and animals interact for them to be called a community, or are such combinations of species just temporary coincidences in time and space? To use a human analogy, we would not call the crowd at a football game a community. The term implies an organised group of people who live together over an extended period of time. It implies a more substantive relationship between the individuals than merely their proximity to one another.

The idea of a community is a way to help ecologists and managers deal with the complexity of ecosystems. For example, in Australia's sub-alpine regions, we can find communities that consist of unique mixtures of plants. Several communities include species that are native to the areas where they are found. Because soil, climate, landform and other environmental factors affect the combinations of species that live in each area, we can use this information to classify the communities into similar groups. This classification helps to make environmental management more effective for conservation (Kirkpatrick and Bridle 1999). One of the most widely studied attributes of ecological communities is their species diversity. This is measured chiefly by richness (the number of separate species present) and evenness (the variation in size between the composite populations).

A good example of how understanding communities assists in understanding the complexity of interacting species and environmental factors can be found in studies of the slow regeneration of old beaver ponds in black spruce (*Picea mariana*) forests. Beaver (*Castor canadensis*) dams create a temporary aquatic environment, but their creators ultimately abandon them because silt accumulates. When an abandoned dam is breached and the pond drains, grasses and sedges grow and persist for decades, apparently resisting any further succession of secondary species, including black spruce trees. Black spruce regeneration after a disturbance requires the presence of mycorrhizal fungi, normally found beneath the soil, but the oxygen-starved environments of the beaver ponds do not allow the fungal pores to survive (Moore 1999). These trees can regenerate in the old dam sites only after fresh fungal spores are brought into the area. Recent research suggests that this fungal dispersal, and subsequent black spruce recolonisation, is achieved via the faeces of small mammals such as the red-backed vole *Clethrionomys gapperi* (Terwilliger and Pastor 1999). This is just one of thousands of examples of how ecologists use studies of community interactions to help understand the dynamics of ecosystems.

The question of what constitutes a community poses a real dilemma for ecology. Do ecological communities really exist, or are they simply artefacts of the way we humans view the natural world? How do we know whether the flora and fauna on a tropical island form a real community or are just independent populations that happened to reach the island? How do we know whether a forest is a real community, or just a random collection of trees that is changing more slowly than we can see?

Science normally starts with a *null hypothesis*. That is, we start with a position that is the opposite of any alternative hypothesis that we might propose to explain some natural phenomenon. In the case of ecosystems, the null hypothesis is that the way an ecosystem functions is insensitive to the addition or removal of species. In other words, communities are just coincidental collections of species, and do not reflect the interactions among the populations within them. Ecologists have proposed several alternative hypotheses to describe the way in which interactions among species influence a community or ecosystem (Lawton 1997).

- The *Rivet Hypothesis* likens species to rivets holding together a complex machine. It suggests that all species contribute to ecosystem performance, so the functioning of the ecosystem will be impaired as its rivets (species) fall out.
- The *Redundant Species Hypothesis* suggests that most species are redundant to the functioning of a community or ecosystem. There is a minimal diversity necessary for proper ecosystem functioning, but the huge diversity seen in most communities is not necessary for the system to operate (Lawton and Brown 1993).
- The *Keystone Species Hypothesis*, which follows the redundant species hypothesis, suggests that only a few particular species are necessary to maintain the main functions of an ecosystem (Woodward 1993). Other species can come and go without having a major effect on the system.
- The *Idiosyncratic Response Hypothesis* acknowledges that the roles of individual species are complex and varied (and usually unknown). Although ecosystem function will change when species diversity changes, the magnitude and direction of such change is unpredictable.

7.4 DO ECOLOGICAL COMMUNITIES EXIST?

The notion of *community* is an important simplifying idea in ecology. As we saw above, the term refers to interacting groups of populations that occur together naturally.

The question of whether the population mixtures that we observe are *natural* is extremely contentious. Does a particular collection of species occur because they 'belong' together or are they simply random assemblages? By 'belong' we understand that the species concerned have adapted to have a suite of mutual interactions and dependencies. Either they cannot survive without the other species, or the community itself is more stable than a random collection of species would be.

The above question is far from hypothetical. Much of the conservation debate centres around the question of whether to conserve whole communities or to focus on saving individual species. If communities are simply random associations of species, then we need not preserve communities; instead we could (at least in principle) conserve species by relocating them in areas not needed for other purposes. On the other hand, if ecosystems do contain sets of species that exhibit strong mutual dependencies, then the only way to conserve those species is to set aside representative tracts of land where their natural communities occur.

The sharp end of the above debate rests with classifications of landscape cover types. Suppose that you need to decide which sites in a region to conserve. *Classification* is the process of grouping together sets of similar sites. The assumption is that all sites classified as belonging to the same group are examples of a single kind of community. Now computer algorithms used for classification can form groups as large or small as you like. It simply depends on where you draw the cut-off level when deciding whether sites are similar enough to belong to the same group. The central issue is where to stop the classification algorithm. Taken to its logical extreme, the conservation argument would be that *every* site constitutes a unique group and should be conserved. The exploitation argument would be that all sites belong to a single group, so it does not matter how many sites are destroyed. The inference that would be drawn from this argument is that to preserve the community we need to retain only a single site.

In much of the above discussion on communities, we have concerned ourselves with patterns – especially with the mix of species that compose a community and the way in which species are distributed across a landscape. In at least some ecosystems, the essential feature is not the pattern but the process involved. The most complex and diverse ecosystems are rainforests and coral reefs. In both cases, the complexity that we see is the product of a long process. In rainforests, it can take many centuries to form rich soils and multi-layered canopies that enable so many species to flourish. Likewise, coral reef communities can come into being only when corals build the reefs that provide protected and varied environments for a host of shallow-water species. In this sense, rainforests and coral reefs certainly are much more than simply random assemblages. They are environmental phenomena that emerge out of long-term interactions.

However, we have not yet answered the initial question. Are communities just random collections of species, or are the combinations selected in some way? To answer this question, we must look more closely at the networks formed by the patterns of interactions among groups of species.

7.5 NETWORKS OF INTERACTIONS

7.5.1 Food webs

During the two month period of May and June 1998, sea lions (*Zalophus californianus*) along the Californian coastline suffered a horrifying incident of mass mortality (Scholin et al. 2000). When scientists investigated the deaths, it turned out that the Californian sea lions were victims of a tragic sequence of ecological interactions. First, a bloom of the alga *Pseudo-nitzschia australis* led to the production of large amounts of toxic domoic acid. High levels of this toxin were then absorbed into the tissues of blue mussels (*Mytilus edulis*) and northern anchovy (*Engraulis mordax*). The toxic mussels and anchovies were ultimately consumed by the top-level predators, especially the sea lions.

The death of the Californian sea lions was an example of the fundamental role played in ecology by relationships between predators and their prey. If species A eats species B, and species B eats species C, then this *food chain* creates a network of interactions that links the destinies of all three species. When we look at an entire ecosystem, then the pattern of interactions (who eats whom) creates what is known as a *food web* (Figure 7-5).

It is through food webs that energy flows occur and nutrients are recycled. To understand ecosystems, we need to understand food webs, because trophic relationships (who eats whom) drive ecosystems. In addition to energy and beneficial nutrients, the structure of food webs can determine the effect of toxins in an ecosystem, as we saw above.

Although understanding food webs is fundamental to understanding ecosystems, there is still much debate about the structure and dynamics of food webs. The determination of food chain length – the number of trophic levels in a food chain – is just one example. Traditional ecological theory suggests that ecosystem productivity determines food chain length, as demonstrated by microbial communities in laboratory microcosms (Kaunzinger and Morin 1998). However, recent research in natural ecosystems has revealed that ecosystem size is a key determinant of food chain length, and not productivity. Within natural lake systems, it was found that smaller, nutrient-rich lakes had shorter food chains than larger, less productive lakes (Post et al. 2000). While such findings identify the relationship between food web structure and ecosystem size, the proximate mechanisms that underlie these relationships are still elusive.

7.5.2 Networks

When species come together, they are bound to interact. In earlier chapters, we looked at the ways in which some kinds of interactions, especially predation and competition, affect the population size over time. However, when we come to look at entire communities, then we need to look at patterns of interactions between dozens, if not hundreds, of separate species, not just a single pair.

Networks of interactions are a major source of complexity in ecosystems. To understand the complexity that is inherent in networks of interactions, let's look at how these networks form. Suppose that an ecosystem consists of (say) 100 species. This collection includes 4950 pairs of species and 9900 possible ways for one species to affect another. Of course, in any real ecosystem, not every pair of species will interact directly with one another. Rabbits, for instance, might not interact with sparrows directly, but they would interact with (say) grass, which

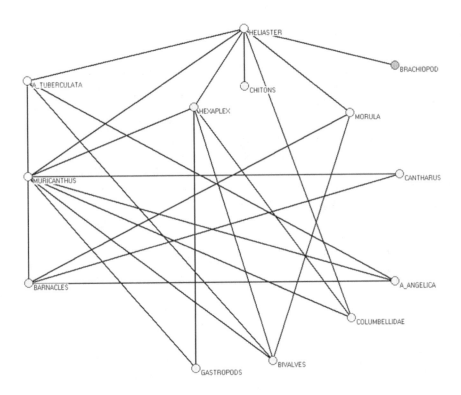

Figure 7-5. Example of a food web. In this case it shows interactions among
species in a marine community (based on data from Paine 1966).

they eat, or with hawks, which eat them. However, rabbits could interact with
sparrows indirectly if sparrows eat the seeds of plants that rabbits eat.

The resulting network would be a network of the kind that we met in Chapter
4. In this case, the nodes of the network are populations, and the edges in the
network are the interactions between populations (Figure 7-6). To look at the
properties of the network, we need to turn it into a model.

One approach is to look at the network as a general competition model. To do
this, we look at each node of the network as representing the size of the population
concerned. Now, each population has an intrinsic growth rate, given by its rate of
reproduction. But each species that interacts with it also affects population growth.
If one population has a positive effect, then it will have a positive coefficient in
the competition model. Rabbits, for instance, would have a positive influence on
the size of a fox population. Foxes, which eat rabbits, would have a negative
influence on the size of a rabbit population. Wherever populations do not interact,
the coefficient in the model would be zero.

To produce such models for real systems is very difficult. As the numbers
given earlier imply, to build a model of a real ecosystem of 100 species, we would
need to determine the values of 10,000 coefficients! These include the 9900
interaction parameters, along with 100 values for the growth rates of individual

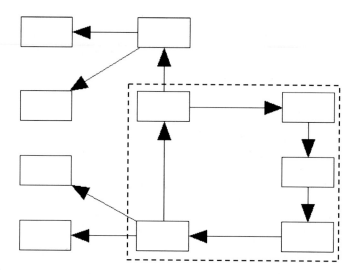

Figure 7-6. An example of a network of interactions. Each box represents a separate population, and the arrows indicate that one population directly affects the other. Examples of such direct effects would include predation, herbivory, and parasitism. The assemblage shown here contains ten populations (boxes) and ten interactions (ten per cent of the maximum possible combinations). In the lower right of the figure, the arrows form a closed circuit, so the five populations enclosed within the dashed box comprise a feedback loop.

populations. To run the model we would also need to provide 100 values for the initial sizes of each of the populations involved.

It is easy to see that ecosystems that are richly interconnected and have high biodiversity, such as rainforests and coral reefs, pose great problems for modellers. It is virtually impossible to determine values of the relevant parameters with any degree of confidence. The famous ecologist Richard Levins used to delight in telling stories about complex chains of interactions between organisms in tropical rainforests. Some of these convoluted interactions confound attempts to model rainforest ecosystems, as well as making management a complex business.

Despite the evident complexity of rainforests, Levins (1977) argued that we can explain many emergent processes, and make useful qualitative predictions, by analysing in qualitative terms the patterns of biotic interactions within an ecosystem. For instance, we know that foxes eat rabbits, even if we do not know the rate. Levins's *loop analysis* consists chiefly of identifying the feedback loops that arise from species interactions. By tracing the sign of each interaction (positive interactions promote population growth; negative interactions inhibit it) we can determine whether feedback is positive or negative. From this sign, we can deduce the emergent properties of the system – that is, whether or not it is stable. Although loop analysis has not figured prominently in subsequent research and management, the lessons it provides have led ecologists to pay more attention to patterns of inter-species interactions, or, in other words, to the structure of food webs.

7.5.3. The paradox of stability

Ecologists have long assumed that the more complex an ecosystem, the more stable it will be. The word 'stability' here means the ability of a community to recover from a disturbance. After being disturbed by a fire (say), the same kind of community will regenerate, and with the same composition as before. By 'complexity' here, people have usually read 'diversity'. As we saw earlier, complexity really arises from interactions. Nevertheless, there is a relationship: the more species there are present, the greater the number of potential interactions.

Charles Elton (1958) identified a potential relationship between the diversity of an ecosystem and its stability. He suggested that decreased diversity would lead to poorer ecosystem functioning and stability. From a thirteen-year study at Cedar Creek Natural History Area in Minnesota, Tilman (1996) presented strong evidence that as species diversity increased, so too did the stability of plant communities. However, there is a discrepancy between the existence of this pattern and the operation of diversity as the mechanism allowing for stability in natural systems (Naeen 1999). Similar research on grasslands in the Western Ghat Mountains of Southern India found mixed results in relation to diversity influencing stability (Sankaran and McNaughton 1999), suggesting that alternative factors (for example, climate) may ultimately influence stability (Naeen 1999). Further, Doak et al. (1998) demonstrated that statistical variations could even be the driving force behind such patterns.

As every first-year science undergraduate knows, correlation does not imply causation. Diversity in ecosystems may cause stability, or stability may allow ecosystems to become diverse. Only experimental manipulation can allow us to distinguish between these possibilities. But ecosystem processes occur on a vast scale in both time and space, making experiments with real systems impractical. Theoretical models, however, have helped researchers to answer this question by providing simple experimental systems that share key properties of real food webs. We shall look at these models and the answers they suggest below.

7.5.4 Stability and food webs

In Chapters 4 and 5, we saw that negative feedback loops promote stability in a dynamic system. Positive feedback loops, on the other hand, create instability. When positive feedback occurs within a network of populations, it usually leads to one or more populations being wiped out. What this means is that if populations are thrown together at random (as might happen in migration onto a volcanic island) then the resulting network of interactions is likely to contain at least one positive feedback loop. This loop will destabilise the system and lead to local extinction of some of the species.

One of the successes of theory in ecology was the demonstration that complexity in an ecosystem (as measured by species richness) does not necessarily imply greater stability. In general, the exact opposite happens: systems that are more complex are less stable (Gardner and Ashby 1970; May 1972, 1974). The underlying cause of this phenomenon is the network of interactions that develops when species occur together. We can understand this issue more clearly by looking at models of networks of interacting populations.

In models of random population networks, instability rules. Random networks that contain more than a handful of species virtually always collapse (Tregonning and Roberts 1979). To understand why collapses occur, look at what happens in a system of interacting populations (Figure 7-6). This model, drawn from work by Newth et al. (2002), examines scenarios in which an ecosystem forms as a random mix of species. The model follows the experimental set-up used by May (1972) and Tregonning and Roberts (1979). That is, it consists of a suite of populations where each population interacts directly with some of the other populations.

In this model ecosystem, feedback loops (Figure 7-6) form inevitably whenever the density of interactions is sufficiently high. The formation of these loops is another side effect of the explosion in connectivity that we discussed in Chapter 4. Make the system rich enough in connections, in inter-species interactions, and the individual connections between pairs of species will link together to form paths by which each species indirectly affects, and is affected by, many other species. Some of these pathways find their way back to their starting point to form a feedback loop. On average, about half of these feedback loops will be negative, and half will be positive. As we saw in Chapter 2, positive feedback loops spiral out of control. What this means for the populations in our model is that, at some point, one or more populations in a positive feedback loop will hit zero. In other words, some species are driven to local extinction.

When we run the above model, the outcome depends on the richness of interactions (Figure 7-7). In this experiment there were initially 25 species, which means that there are 600 (25×24) different pairs of species. If the number of pairs that actually interact is low as a percentage of this total, then virtually no species are lost. In the case shown (one per cent of all pairs), this means that the number of species in the system continually grows as more and more invaders enter the system. However, when the percentage is substantial (for example, 30 per cent), then positive feedback loops form (Figure 7-6), and many species are quickly driven to extinction. In the scenario shown in Figure 7-7, with 30 per cent of all pairs interacting, the model immediately crashed from a seed mix of 25 species down to just six species. However, those survivors formed a stable core (Figure 7-8). The system then oscillated between stability and instability. Subsequent invasions led to growth, and occasional crashes, until the ecosystem achieved a more or less steady state with about 15 species. A higher percentage of interactions made it progressively harder to find stable combinations of species, and crashes were much more common. So in the case where all species interacted, then no more than ten species were able to coexist at any time.

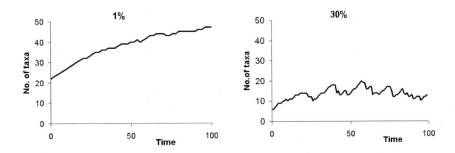

Figure 7-7. Changes in the number of populations (taxa) within a random network of interacting populations. The mixtures of taxa are formed at random, as in the previous figure. The populations interact according to a generalised competition model. The percentages above each graph indicate the proportion of all possible interactions that actually occur. The sign of the interaction (positive or negative) between each pair of taxa is randomly assigned. At each stage, the model was run for 100 iterations, and any species with a zero population was removed. The numbers of survivors are plotted in the Figure. In the scenario shown, the initial assemblage was 25 species, each of 1000 individuals. At the conclusion of each stage, the surviving species, plus one new 'invader' species, formed the species assemblage for the next stage. (After Newth et al. 2002.)

Finally, the most significant result that arises from the above studies is that at the end of the experiment, the network of species interactions is no longer random (Newth et al. 2002). No positive feedback loops are left; only negative feedback loops, which stabilise the system. Also, the density of connections (about 5-20 per cent) is usually lower than the critical density needed for a random network to become fully connected and to form positive feedback loops spontaneously. The pattern of connections changes too. Rather than being a random network (for example, Figure 4-3a), order has developed. For instance, some populations tend to have more interactions than others. So the pattern of connections looks more like a small world (Figure 4-3e) or a scale-free network (Figure 4-3f).

Studies of real ecosystems tell a story that is consistent with the above experiments (for example, Paine 1966; Pimm 1982; Solé and Montoya 2001; Montoya and Solé 2002). They have established that the density of connections between species is never very high, the range being 5-25 per cent. In most ecosystems, the density is below the critical level, above which the formation of positive feedback loops becomes inevitable.

The importance of these results is that they confirm the existence of 'ecological communities'. The experiments imply that mature ecosystems are not random collections of species at all. They are the result of the evolution of network relationships that result in stable structures. Because connectivity is

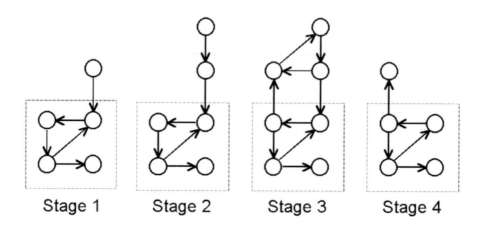

Stage 1   Stage 2   Stage 3   Stage 4

Figure 7-8. Formation of a stable core in a random dynamic network. At each stage, a new node is added to the core (shown here as the nodes in the box). Over time, a series of additions and minor collapses (Stage 4) lead to larger stable networks.

relatively low, the communities are capable of growing into very diverse ecosystems. The experimental results imply that the dynamics appear to be self-selecting, so the communities that we see in nature can be seen as attractors of the process. Over long periods, it is likely that the composition is reinforced by populations within particular communities further evolving in ways that enhance the stability of particular network structures.

The experiments also offer theoretical confirmation of the importance of *keystone species*. These species are thought to play an essential role in holding particular communities together. Remove the keystone species and the community may collapse. The model readily explains how this phenomenon could occur. Keystone species are the most highly connected in the network. Remove them and the entire pattern of connections falls apart. In the next chapter, we shall look at how these large-scale ecological processes may interact with natural selection and other genetic effects to generate patterns in the diversity of life and ecosystems.

CHAPTER 8

GENETICS AND ADAPTATION IN LANDSCAPES

*Tiny Argentine ants (*Linepithema humile*) are highly successful invaders because the colonists' low genetic diversity inhibits aggression between nests. Here, they co-operate to drive a native bulldog ant (*Myrmecia sp.*) away from a food source in southern Australia.*

abundant due to chance (*genetic drift*). Eventually, only one version of each allele persists as the other version has drifted to extinction. So in this model, the distribution of the trait has a single peak and slowly becomes narrower. We call the movement of genes within and between populations *gene flow*.

8.1.1 Genetic trade-offs

Which would you rather be, a lion or a zebra? A cat or a mouse? Most people, when asked this question, would opt for the carnivore. This is not really surprising: who wants to be eaten? More generally, we might think it is an advantage to be large, swift, strong, omnivorous, fast-growing, and able to tolerate any environmental conditions. So why do we not see all of these qualities in a single species? One reason is that every attribute comes at a cost. To be good at one thing, you have to sacrifice something else.

These genetic trade-offs are found everywhere in nature. One classic example concerns reproduction. At one extreme are species such as locusts that produce vast numbers of offspring and leave them to take their chances. Although most will quickly perish, some will survive to reproduce in their turn. At the other extreme are species (including mammals and birds) that produce just a few offspring, but devote time and energy to rearing them, so that a relatively high proportion survives to reproduce. Both strategies work, but no species can achieve both extremes at the same time. There are many other examples of genetic trade-offs. In Chapter 6 we saw that mangrove species have to make a genetic trade-off between growth rate and salt tolerance. Almost any adaptation that enables a species to cope with extremes involves a similar kind of trade-off.

A poignant example of the evolutionary dilemma posed by genetic trade-offs can be seen in the cheetah. Cheetahs are the world's fastest mammal, reaching speeds of up to 104 kilometres per hour (Sharp 1997). Cheetahs have evolved to be so fast by preying on small, swift antelopes. However, being so extraordinarily fast also means being light-framed. Consequently, they are unable to bring down any prey much larger than a gazelle, and are easily driven away by more robust predators such as lions. They cannot even defend their own cubs, who are commonly eaten by lions. Models suggest that as a result, cheetahs will rapidly become extinct wherever lions are abundant (Kelly and Durant 2000).

One of the most important genetic trade-offs is between generalization and specialization (Futuyma and Moreno 1988). Ecological generalists are species that survive and breed successfully in a wide range of conditions. However, in any specific situation, generalists may be out-competed by specialists. A Swiss Army knife is handy when you are on holiday, but in normal life you are likely to prefer separate implements for knives, forks, corkscrews, and so on. Similarly, species that possess the diverse adaptations required to tolerate a wide range of conditions are unlikely to out-compete species that specialise in one particular environment. Generalists thus predominate in unpredictable environments while stable environments promote specialists. This observation may have some powerful implications for macroscopic evolution, as we shall see later.

8.2 GENETICS IN HETEROGENEOUS LANDSCAPES

Variation in conditions across landscapes leads to variation in selection pressures. For example, hotter areas promote adaptations for heat loss; cooler areas promote well-insulated animals. These variations can be influential at a very local spatial scale. For an animal living on a hill, a gene that facilitates hill-climbing by enhancing the leg muscles might be useful, but the gene could be wasted on the same animal living on flat territory just a few metres away. Moreover, such a gene could be maladaptive if it results in the animal diverting resources to muscle instead of, for example, to brain. Environmental heterogeneity thus slows natural selection by creating varied and conflicting selection pressures.

This effect appears to explain one of the great mysteries of evolutionary theory, the *paradox of the lek*. A lek is an area where male animals congregate to display before females. Males often possess extravagant ornaments, such as the tail of the male lyrebird (Figure 8-2), and expend much time and energy displaying before females, who choose their mates carefully. It is thought that these traits evolve because the genes of males who attract larger numbers of mates will be disproportionately represented in the next generation. Likewise, the genes of females who choose popular males will be disproportionately represented in proceeding generations because their male offspring may inherit the desirable displays of the father and the wise preferences of the mother. This creates an evolutionary positive feedback loop that progressively exaggerates male traits and female preferences (*runaway sexual selection*) (Fisher 1930).

Figure 8-2. A male lyrebird sings, dances and displays his enormous tail as a mating advertisement in a rainforest in south-eastern Australia. Such extreme traits are thought to evolve through runaway sexual selection.

But therein lies a paradox: if males with desirable displays have more offspring, then desirable displays should soon become ubiquitous, leaving no grounds for female choice, thereby rendering the displays meaningless. Models suggest that as a result, if choosing between males requires even a small effort, female choice and male displays both become maladaptive (Bulmer 1989). It was unclear, then, why showy males and choosy females are so prevalent among animals, and why variation in displays and preferences persists.

The answer, it seems, may lie in spatial variation. Using a spatially explicit model, Day (2000) demonstrated that varied and extreme displays and preferences could be maintained indefinitely even when female choice was quite costly. Such ongoing evolution occurred only, however, when the male traits that were most adaptive varied across the spatial environment. This spatial variation maintained conflicting selection at different locations, so that runaway sexual selection occurred continually, but tugged the population in several different directions.

Similar spatial effects of mate choice on genetic diversity can be seen in the cellular automaton model that we looked at previously (Figure 8-1), if we suppose that individuals choose their mates based on a genetically encoded preference for the colour shade trait. Starting from a random state as before, the model quickly evolves into a patchwork of discrete areas, each highly uniform and distinct from its surroundings (Figure 8-3). Each patch is an area where runaway sexual selection has eliminated diversity, but sexual selection has favoured different colours in each location.

Figure 8-3. Runaway sexual selection of a colour shade trait and preference in a two-dimensional cellular automaton landscape where cells exchange genetic material with their neighbours. Time flows from left to right. Initially, genetic variation is randomly distributed (far left panel), but as time passes, discrete patches of uniform colour evolve.

8.2.1 Adaptation on a gradient

One of the most pervasive landscape patterns is the environmental gradient. We encountered gradients in Chapter 6 with regard to salinity in an estuary. Likewise, gradients in temperature, pressure, light, moisture, and the concentration of chemicals are common in landscapes. These abiotic factors can also generate gradients in the abundance of living organisms; that is, ecological gradients. As we saw in Chapters 6 and 7, studies of species distributions show that inter-species competition influences species distributions. Likewise, gradients can play an important role in the adaptation of species to different conditions. The interactions between these factors can produce complex and surprising spatial patterns.

Let's take a simple example. What would happen if a vast unoccupied region confronted a number of competing populations, but none of them could exploit it because some environmental parameter was too extreme for them to tolerate? This could occur, for example, among mangroves in a salt marsh, trees on a mountainside, grasses at the edge of a desert, or animals in sub-polar regions. One way to study this question is to simulate a simple system. We can place a number of species on a simulated landscape with an environmental gradient running across it. Suppose that all the species present are initially adapted only to low values of the gradient. None of them can survive above a certain upper limit. This means that all the populations are restricted to live in a narrow band at one end of the environmental gradient. However, at any time there may be offspring with mutations that enable them to cope with greater values of the gradient.

When we run this model, the results show a dramatic effect of chance historical events (Figure 8-4). Sooner or later, one species gets an adaptation that enables a few individuals to survive in the empty region. These pioneers then spread and diversify. Usually, the population continues to adapt to the more extreme conditions until the species occupies the entire region, shutting the other species out. The genetic adaptation enhances the effect of competitive exclusion, making it much more pronounced than would normally result from simple competition.

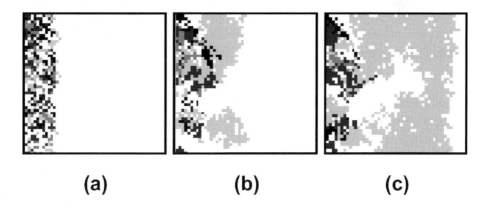

(a)   **(b)**   **(c)**

Figure 8-4. The chance acquisition of an adaptation influences species distributions on a habitat gradient. In this experiment, an environmental gradient runs across the landscape from left to right. (a) Several competing species are initially confined to the lower end of the gradient because they are unable to tolerate higher values. (b) A tolerance-increasing mutation appears and spreads through one species, allowing it to spread into areas higher in the gradient. (c) This species consolidates its dominance of the upper gradient even after other species adapt to it as well.

8.2.2 Fragmentation and drift

One of the assumptions of the Hardy-Weinberg law is that populations are large. In fact, literally speaking, the law assumes that populations are infinite, but violations of this assumption are unimportant unless populations are small and/or the time-scale in which we are interested is long. Chance becomes more important to the fate of genes when populations are small. Because mutation is rare, most of the activity in changing gene frequencies at any time occurs among variations that arose previously, rather than involving novel mutations. In each generation, even when populations are not under selection, the frequency of different genes changes very slightly due to chance. If we plotted the frequency of a neutral gene over generations on a graph, then the Hardy-Weinberg Law predicts that the line should be perfectly horizontal. In reality however, although each point will be near the preceding one, it will not be exactly on par, and over many generations the line may waver far from its original value. This behaviour is called a *random walk* and in genetics it is known as *drift*.

Each neutral allele performs a random walk between disappearing altogether (*extinction*) and becoming the only variant present in the population (*fixation*). In each generation, the allele moves one step either away from extinction and towards fixation, or towards extinction and away from fixation. The number of steps between the two possible fates is proportional to the size of the population. Consequently, in a small population, every gene is always much closer to chance extinction or fixation than in a large one.

Landscapes enter this picture because fragmentation can effectively break down the connectivity within a population and reduce it to a set of small, isolated subgroups. Before going any further, it is important to make clear exactly what the term *connected* means in this context. It means that populations at different locations in a landscape can interbreed with each other (i.e. they can share genetic information). Processes such as seed dispersal and pollination therefore determine connectivity and are essential to maintain genetic homogeneity within populations.

The key question here is when and how the genetic connectivity in a population breaks down. To simulate the process, we can use a cellular automaton model to represent the landscape as a square grid of cells. A random subset of the cells is occupied by organisms (*active*); other cells represent unused territory. The model does not specify the reasons why particular territory is unused, but there are many possibilities, such as intolerable physical conditions, or the presence of ecologically incompatible species.

The neighbourhood of each cell's inhabitants is the set of all surrounding cells inhabited by organisms that can breed with them. For plants the radius of the neighbourhood might be the distance that pollen can spread. For animals it might be distance that they travel from the centre of their home territory. A population is effectively 'connected' if genetic material can flow right across the entire territory it occupies. This means that between any two points in the territory, there must be a sequence of directly connected sites by which genetic information can flow.

As we saw in Chapter 3, a phase change in landscape connectivity occurs as the proportion of occupied sites in a landscape increases. The exact value of the critical point depends on the size of the neighbourhood. If the connectivity between sites falls below this critical level, then a regional population effectively breaks up into genetically isolated subpopulations. These isolated subpopulations will be relatively small. Consequently, each gene within them will be relatively close to extinction through drift. A fragmented population therefore tends to have much lower genetic diversity than a well-connected one, because the variations within it go extinct more rapidly.

Loss of genetic diversity within isolated subpopulations can cause severe problems. Many gene variations are masked when an alternative version is also present, requiring two copies of the variant for any effect to be seen. These variants are said to be *recessive*. Some gene variations are harmful, and therefore removed by natural selection. However, because recessive variations have no effect unless two copies are present, harmful recessive variations are most often carried and passed on by healthy individuals. Consequently, although such genes are gradually removed by natural selection, the efficiency of the process is low and declines with the recessive gene's frequency[2]. The result is that numerous harmful recessive variations are present in virtually all populations, but each such variation is normally rare.

When populations become fragmented and genetic diversity is lost, harmful recessive mutations often become common or ubiquitous within small subpopulations purely by chance. At worst, this process can lead to *mutational meltdown*, where many moderately harmful variations that would normally be rare become ubiquitous and interact to drastically reduce the fitness of the population as a whole, potentially driving it to extinction (Lynch et al. 1995). Cheetahs, for example, are almost completely genetically uniform. The consequences include high incidences of abnormal sperm, juvenile mortality, and mortality from normally minor infections (O'Brien et al. 1985). The same effects may generally increase the vulnerability to extinction of populations living on small, isolated islands (Frankham 1998).

However, loss of genetic diversity is not the only consequence of fragmentation. Although fragmentation reduces genetic diversity at a local scale (within small, isolated subpopulations), on a larger scale (within whole landscapes over long periods) it promotes evolutionary divergence. The reason is that although each allele is close to extinction in a small population, it is also close to fixation. In a fragmented population, each new mutation therefore has a much greater probability of fixation than in a connected population. Each gene therefore takes its own evolutionary path within each sub-population, leading to ongoing differentiation. The result is that fragmentation enhances genetic differentiation despite reducing genetic diversity locally. Given sufficient time, this process can lead to the evolution of new species (*speciation*).

---

[2]The Hardy-Weinberg Law implies that the frequency of expression of a recessive allele is the square of the allele's frequency. For example, a recessive allele that has a frequency of 70 per cent is expressed in 49 per cent of individuals, but a recessive allele that has a frequency of 35 per cent is expressed in only 12 per cent of individuals: so in this case, halving the gene's frequency reduces the selection pressure by about 75 per cent.

We can again illustrate this effect of landscape connectivity on the differentiation of a gene by using a cellular automaton model (Figure 8-5). Suppose that we can represent this gene by a real number G. The number G might represent (say) height, or growth rate, or amount of skin pigmentation. Then each organism (each cell in the model landscape) will contain a local value of G. Each cycle of the model represents a turnover of generations; random perturbations of a cell's G mimic the effects of mutation, and sexual reproduction randomly replaces one cell's G with the G of an adjacent cell.

We start by randomly setting each cell's G value. If the landscape is well-connected, then the trait eventually becomes uniform everywhere. Gradually, the alternative G versions become more or less abundant due to genetic drift. As we saw in earlier chapters, positive feedback plays an important role in this process, with common values growing at the expense of rare ones. After a very large number of generations, only one version persists as the other versions have drifted to extinction. New mutations virtually never become fixed because they are so rare relative to their competitors.

As we saw above, if the connectivity between sites provided by dispersal falls below a critical level, the regional population effectively breaks up into genetically isolated subpopulations. As before, each of the alternative alleles in these subpopulations drifts between extinction and fixation. However, the isolation of each subpopulation means that its dynamics are decoupled from those of the subpopulations around it: drift can go in different directions in different subpopulations. Because the subpopulations are much smaller than the fully connected population, genetic drift proceeds quickly and new mutations are closer to fixation from the beginning. Drift may remove different versions of the same gene in different subpopulations, causing the subpopulations to diversify over time. Although each subpopulation has lower genetic diversity, in the long term evolutionary diversification is greater in the fragmented populations.

The above effects are apparent in simulation results (Figure 8-5). In a fully connected landscape, reproduction functions as a spatial filter. It slows genetic drift in uniform populations and causes heterogeneous populations to converge. However, if connectivity falls below the critical level, then genetic drift proceeds unimpeded in initially uniform populations, and heterogeneous populations do not converge but continue to drift apart. Note that the critical region in these simulations (40-60 per cent coverage) is specific to the neighbourhood function used and varies according to the pattern of dispersal. These results imply that the genetic makeup of a population is highly sensitive to changes in landscape connectivity.

Daniels et al. (2000) used a spatially explicit individual-based model to look at the factors affecting genetic diversity in fragmented populations of endangered red-cockaded woodpeckers. Incorporating realistic limits to migration and dispersal in space, as well as the natural social behaviour of the birds, allowed the researchers to duplicate in the model the patterns of genetic diversity and inbreeding seen in nature.

These effects have also been documented at finer scales in both time and space. Roadways, for example, form a significant barrier to migration of small

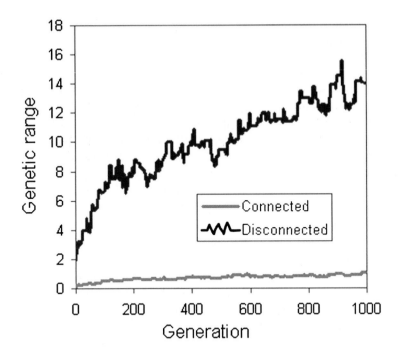

Figure 8-5. Simulated genetic drift in a landscape. The figure shows the range of gene values (*G*), after 10,000 'generations' of a population that initially is: (*lower line*) homogeneous (*G* = 0 everywhere); and (*upper line*) heterogeneous (-100 < 100), in response to the proportion *P* of active sites. In both cases, the range 0.4 < *P* < 0.6 forms a critical region.

mammals. This has led to detectable genetic fragmentation of bank vole (*Clethrionomys glareolus*) populations in southern Germany and Switzerland (Gerlach and Musolf 2000).

8.2.3 Friends and relations

One of the most active areas of evolutionary research over the last few decades has been the study of how co-operative behaviour can emerge. Co-operation involves self-restraint or self-sacrifice for the benefit of another. Co-operative behaviour is a puzzle because, as discussed above, natural selection should lead to genes behaving selfishly. Moreover, natural selection has no foresight: what is selected is whatever happens to be adaptive right now for currently existing individuals, not what will be adaptive for all individuals in the long run.

Genes that cause individuals to provide resources to their own kin are often selected for because those kin carry many of the same genes (*kin selection*). For instance, by supporting their queen, sterile worker bees promote the survival of their own genes. Kin selection is extremely powerful, particularly where offspring are concerned. Mammals gestate and nurse their babies for months. Leeches brood their eggs, guard their hatchlings, and sometimes even hunt prey for their offspring (Kutschera and Wirtz 2001). Perhaps the most extreme maternal sacrifice, however, is seen in the spider *Amaurobius ferox*. In this species, a complex series of communication signals is exchanged between mother and offspring over several hours. These culminate in the mother performing soliciting behaviours which stimulate her young to devour her (Kim and Horel 1998).

Co-operation also occurs between unrelated individuals in the form of reciprocity. Reciprocity is trade: one organism makes a small sacrifice in order to provide a large benefit to another, who does the same in return. Vampire bats, for example, live on the blood of large mammals. They cannot survive more than a few days without feeding, but if they do feed, they will consume far more blood than they need. After a night's foraging, upon returning to their roost vampires who have fed will commonly donate blood to those who have not, helping them to survive the day at a small cost to themselves (Wilkinson 1984).

Reciprocity is often studied formally using the famous 'Prisoner's Dilemma' game. Two prisoners are each offered freedom if they incriminate the other (this is the temptation to behave selfishly). If neither talks (i.e., the prisoners co-operate), they each suffer a small punishment. If both talk, then both receive the maximum sentence. If only one talks then he goes free, while the other receives the maximum sentence.

When games like this are played repeatedly, as among the vampire bats, reciprocity increases everyone's fitness. However, the individual that receives the biggest advantage of all is one who takes the benefit of others' help, but refuses to pay any costs. An individual with such a mutation could out-compete its co-operative peers, and so the co-operative behaviour would soon disappear. The problem with evolving a co-operative group is how to recognise and exclude such cheats (Dawkins 1976). This problem probably explains why reciprocity appears to be very rare in nature: unless co-operators can exclude cheats, cheats will out-compete co-operators; consequently, there will soon be no co-operators left.

Incorporating landscapes and social networks into models can make co-operation and self-restraint more likely to emerge for several reasons. First, existence in a spatial landscape or social environment implies that individuals are likely to encounter one another repeatedly, allowing them to develop sustained social relationships. In terms of reciprocity, this factor potentially allows them to recognise and avoid 'cheats'. Secondly, nearby individuals will often be kin, so it may pay genetically to be less than ruthless. Both of these effects contribute to clustering of co-operative individuals in space, reducing their vulnerability to exploitation.

The effect of spatial geometry on the evolution of co-operation was dramatically illustrated in a study by Nowak and May (1992). The authors studied the evolutionary dynamics of a Prisoner's Dilemma where the players' strategies were genetically determined. There were only two strategies: always co-operate or never co-operate. In non-spatial versions of the game, co-operation would disappear rapidly. However, when the players were placed in a two-dimensional spatial array, kin selection generated chaotically changing fractal patterns of co-operative and cheating behaviour which lasted indefinitely. Models that allow more sophisticated game-play strategies generate co-operation even more easily (Nakamaru et al. 1997, 1998; Brauchli et al. 1999).

Another promising approach looks at the evolution of co-operation in marginal habitats (Avrilés 1999). In comfortable environments, population pressures mean that competition between individuals soon becomes a major evolutionary factor. In habitat that is too poor, populations cannot become established. But at the margins between these habitats, where the species can just barely persist, internal competition between individuals may be less important for fitness than simply surviving the external conditions. Consequently, selection may favour co-operative behaviour in marginal habitats, where individuals band together to survive external stressors.

The conflict between group and individual selection and its role in success across landscapes is vividly illustrated by the case of the invasive Argentine ant (*Linepithema humile*). These are tiny brownish-black ants that you will find in buildings, parks and gardens all over the world (see the photo at the start of this chapter). The reason for their extraordinary success as colonists lies in a quirk of genetics (Suarez et al. 1999; Tsutsui and Case 2001). Normally, different ant nests expend considerable effort in competing aggressively with one another, recognising foreigners by subtle variations in a genetically determined chemical signature. In their native land, Argentines also do this, and they have not attained any dominance over the many other ant species present. However, because the original Argentine invasions on other continents began with only a few nests, the Argentines that have become ubiquitous worldwide are all very closely related. They are so genetically similar that they all share the same chemical signature. Consequently, they all regard one another as nest-mates, and co-operate instead of competing. This factor probably explains their success: while native ant species are preoccupied with internal warfare, the Argentines band together to defeat them. However, in the long-term, the Argentine co-operative behaviour seems unlikely to be stable. Genetic drift will diversify the populations, and competition between nests will become increasingly important to survival as the landscape is filled, ultimately causing selection favouring the more competitive nests.

More speculatively, some authors claim that co-operative behaviour can evolve to stabilise population dynamics (because individuals that destabilise their own populations die with the group). For instance, in an agent-based model of animals foraging on a renewable resource in a patchy landscape, Pepper (2000) found that restrained behaviour was favoured. When greedy individuals found a patch of food, they would eat everything in sight. However, the resulting local population explosion soon exhausted the plants in the patch, leading to local extinction. This effect is akin to the dilemma that diseases and parasites face: high virulence allows them to reproduce rapidly, but may also kill their hosts before they can spread. In contrast, restrained feeders did not gain as much energy from any patch they occupied and were quickly out-competed by greedy individuals wherever they occurred together. However, because the environment was patchy, there were areas where they did not have to compete. Populations of restrained feeders were able to survive in these patches because they maintained a sustainable harvest.

At an even broader scale, some researchers argue that ecosystems evolve towards a stable state, with ecosystems that contain destabilizing positive feedback processes gradually being replaced by more stable ones. The strongest extension of this view is the Gaia hypothesis, which argues that the entire biosphere evolves to a self-maintaining state. We will look at this idea, and the reasoning behind it, in more detail in Chapter 9.

8.3 CATASTROPHES, CRITICALITY AND MACROEVOLUTION

The story of planet Earth is one of dramatic changes brought on by interactions between life and the physical environment. For example, it is thought that in life's early history, conversion of hydrogen to methane by ancient bacteria created a greenhouse effect sufficient to maintain widespread liquid water despite the weak light of the sun (Kasting et al. 2001). Around 2.1-2.4 billion years ago, the spread of cyanobacteria caused a massive rise in atmospheric oxygen, transforming the planet's atmosphere from a neutral or reducing state to an oxidizing one (Wiechert 2002). In turn, rising oxygen created a blanket of atmospheric ozone, protecting the planet from DNA-damaging ultraviolet rays, and thereby allowing life to spread across the planet's surface (Knoll 2003). We can imagine that each of these events, by enormously altering the environment, will have driven many species to extinction, and created opportunities for the evolution of many more.

Disasters form another source of change. On a global scale these include volcanic eruptions, earthquakes and asteroid impacts. Disasters that appear localised can lead to wider or even global consequences, such as tsunamis, dust storms, and climate change. As we saw in Chapter 7 with the keystone species concept, disruption to ecosystems could generate ripples of extinction even among species not directly affected by the catastrophe. Such a large and widespread loss of species is called a mass extinction. The species that survive mass extinctions are not a random sample; instead, they tend to be widespread, and ecological and environmental generalists (Erwin 1998).

Since the evolution of complex multicellular life forms, there have been five known events in which over 50 per cent of species became extinct (Raup and Jablonski 1986). One of the best-studied is the mass extinction which ended the Cretaceous and began the Tertiary period with the disappearance of the dinosaurs. In 1980, Luis Alvarez and his colleagues demonstrated that an asteroid impact was associated with this extinction (Alvarez et al. 1980). Their evidence was a thin layer in the Earth's crust that is rich in the metal iridium. This layer coincided precisely with the Cretaceous-Tertiary (K-T) boundary. Subsequent research has identified that this iridium layer occurs worldwide. It seems to have resulted from an impact on the Yucatan Peninsula in Central America. More recent evidence suggests that catastrophes have caused extinction events throughout the history of life on Earth (Hallam and Wignall 1997).

Some authors argue that mass extinctions may also self-organise without a large external cause: ecosystems may become more brittle over time until a disturbance causes an extinction event (Solé and Manrubia 1996). One such extinction could cause others, generating an avalanche of extinctions. Because ecosystems do not have clear boundaries, the largest of such events could percolate to the global scale. Such increasing brittleness of ecosystems could occur if the evolution of new species destabilises ecological networks, or if (as suggested earlier) ecological interactions become more specialised over time in stable environments. Thus, according to this hypothesis, ecosystems continually evolve towards a critical state. This phenomenon is called *self-organised criticality*, and one of its key features is that it produces a power law distribution of event sizes. Proponents of the theory point out that the size distribution of mass extinction events does indeed appear to follow a power law (Solé et al. 1997). However, this finding could have other explanations (Newman 1997). In addition, the ecological assumptions required by this theory are restrictive. Overall, whether self-organised criticality plays any part in mass extinctions remains uncertain.

Early Darwinian theory had little to say about the rate of evolutionary change produced by natural selection. In the late 20th century, mounting fossil evidence suggested that the rate of evolution may have varied widely through history. Eldredge and Gould (1972) argued that a common cycle is evident throughout evolutionary history; occasional mass extinctions are followed by bursts of sudden evolutionary novelty (Raup 1986, Raup and Jablonski 1986, Kauffman and Walliser 1990), followed by long periods of stasis which are eventually broken again by mass extinction. They termed this pattern *punctuated equilibrium* and argued that it reflects an evolutionary history dominated by constraints imposed by existing adaptations to stable environments. Both mass extinctions and evolutionary diversification after them are well-established. However, whether the apparent stasis of the fossil record over long periods reflects a real equilibrium state of evolution, or whether it merely reflects the incompleteness of the record, is still widely questioned.

Gould (2002) extended the punctuated equilibrium argument even further, arguing that much of the significant activity in evolution is caused by selection between species rather than individuals. If, for most of evolutionary history, species remain fairly stable because constraints prevent individual-level selection from acting effectively, then individual-level selection may become less important to a species' fate than its ecological interactions with other species. Unstable ecological interactions, such as competition, can be seen as imposing a form of species-level selection: the superior competitor ultimately drives the inferior one to extinction. Individual-level selection may be powerless to prevent this result because of the evolutionary trade-offs the species has already made. We saw this effect in the example of cheetahs above. However, the prevalence of such evolutionary constraints remains uncertain.

8.3.1 Landscape phases and the origin of species

We can briefly summarize the role of landscapes in macroscopic evolution as follows. For most of the time the system – that is, populations of plants and animals in their environment – remains undisturbed. In this situation, individual species exist in either of two states: connected or fragmented. Because the transition between these two states occurs across a critical threshold, populations are unlikely to be in an intermediate state; instead, we expect them to be either overwhelmingly connected, or overwhelmingly fragmented.

For species that consist of a single, connected population, genetic information is constantly being circulated throughout the entire population. As we saw earlier, the effect of this constant mixing is to inhibit diversification (Figure 8-5) and slow genetic drift. Natural selection gradually increases the average fitness of the population, but recombination continually breaks down associations between genes, preventing different parts of the population from adapting to different local conditions.

There is now a very large literature on how spatial distributions affect *speciation*, the process by which new species arise. A species is defined by reproductive isolation: that is, gene flow between different species is insignificant or non-existent, even when the populations are not spatially isolated (Mayr 1942).

The simplest way to stop gene flow between populations is landscape fragmentation (Mayr 1942). When a species is broken down into isolated populations without gene flow, each population proceeds along a separate evolutionary path. Neutral variations diversify. Because the strength and direction of selection varies from place to place, selection can accelerate and magnify genetic divergence between subpopulations. Genes may arise in one population that are incompatible with genes that arise in the other, causing sterility, inviability or maladaptation of potential hybrids. Sexual selection can also contribute by causing the mate choice and fertilization systems of the two populations to diverge.

Geographic separation does not need to be absolute or very long-lasting to contribute to speciation. Speciation can occur in connected populations when there is strong selection in different directions acting on different parts of a population. One of the most likely scenarios for this is the presence of an environmental gradient, where gene flow between the speciating groups is reduced by both natural selection and spatial associations between genes. In this situation, which was modelled by Doebili and Deickmann (2003), selection can rapidly lead to the formation of two discrete, well-defined species living alongside one another.

A common occurrence is that two such adjacent populations are able to interbreed but with only partial success. Such populations, which are intermediate stages in speciation, may have arisen on an environmental gradient as described above, or they may be the descendents of populations that have been fragmented at some time in the past and have later migrated into contact. In this situation, the populations often do not merge. Instead, hybrids are confined to a narrow, linear interface between the two populations called a hybrid zone (Barton 1979). The reason the populations remain stable and do not merge is that genes derived from each population are disadvantaged on the side where they are in the minority. Over time, further isolating mechanisms may accumulate in these populations, so that eventually they become distinct species.

Landscape fragmentation at finer spatial scales can also enhance diversification. Mosaics are a common landscape pattern where patches of different habitat types are interspersed. Cain et al. (1999) examined how selection could lead to speciation via the divergence of mate choice systems in mosaics where two subspecies hybridize. Sadedin and Littlejohn (2003) developed a spatially explicit agent-based model of this speciation process. The model showed that the likelihood of speciation was governed by a critical threshold whose order parameter was the level of hybrid disadvantage. If the level of hybrid disadvantage was lower than a certain value, then speciation never occurred. Above this range, there was a narrow band in which where the outcome was uncertain, but above this band, speciation always occurred. The results also showed that habitat heterogeneity decreased the threshold level of hybrid disadvantage needed for speciation. The striking similarity of this effect to percolation processes (Chapter 3, Section 5) suggests that speciation in these hybrid zones may involve a connectivity avalanche.

Overall, this evidence suggests that phase changes in landscape connectivity at many scales are potentially a powerful source of evolutionary change. There is also evidence that such phase changes occur in real landscapes. In central Australia, for instance, the varying availability of lakes, and other bodies of water limits the distribution of water birds, such as ducks. Although many species are widespread, Roshier et al. (2001) found that the centre of the continent showed two distinct phases. Suppose that small birds can fly a reasonable distance to reach water, say 100-200 kilometres. Then in years of high rainfall, the entire centre is effectively connected. That is, birds could comfortably fly anywhere by moving between sources of water. However, in years of drought, many water bodies dry up completely. As far as the birds are concerned, their movements are effectively restricted to small isolated regions. Presumably the variation in connectivity of water bodies means that the extent of interbreeding also varies from season to season in response to rainfall. The long-term impact of these variations on the bird populations is not known. Nevertheless, they do show that real changes in landscape connectivity do regularly occur. For less mobile organisms, the effects are likely to be much more pronounced even locally.

8.3.2 Cataclysms and punctuated equilibria

The slow, steady accumulation of genetic changes is often interrupted by disturbances. The history of life is peppered with cataclysms, both great and small. Most discussion of cataclysms has focused on the impact of comets and other events with the potential to disrupt the entire planet. However the biosphere is continually subjected to cataclysms of all sizes. Great events, such as the impact of a comet, are as rare as their effects are vast. Smaller events are more common. So common are small disturbances that every year the Earth's surface is marked by thousands of fires, storms, volcanic eruptions, and other events.

The most important effect of any cataclysm is to clear large tracts of land at a single stroke. In doing so, the cataclysm plunges an ecosystem into a different phase. Suddenly the normal rules no longer apply. Formerly connected populations may be fragmented, while fragmented populations may recombine.

Landscape fragmentation may play a role in the observed pattern of punctuated equilibrium, where long periods of stasis are followed by mass extinctions followed by sudden bursts of evolutionary novelty. Gould (2002) argued that mass extinctions, by releasing evolutionary constraints on surviving species, facilitate diversification of the survivors.

One possible mechanism for this release of evolutionary constraints is the spatial suppression of competition. In Chapter 6 we saw how pollen zones reveal an ecological pattern resembling punctuated equilibrium in the recent history of forest communities. This pattern arises because when dispersal occurs locally, well-established species suppress invading newcomers until a major fire clears the landscape, triggering population explosions of the invaders (Green 1982, 1987). The parallel between these processes occurring on vastly different scales suggests that common processes underlie both pollen zones (on the scale of millennia) and punctuated equilibria (on the scale of geological eras). New species that have arisen in fragmented populations may be unable to spread in landscapes under normal conditions because of the spatial dominance of their competitors. By clearing large tracts of land, cataclysms remove the spatial advantage of established species. Thus, the explosions of novelty seen after mass extinctions may not be due solely to innovation after the cataclysm; instead, they may represent explosions of species that were already present in small numbers, but had been suppressed and confined spatially.

Another possibility is that punctuated equilibrium is at least partly driven by the pervasive evolutionary trade-off between specialization and generalization. The species that are driven to extinction by cataclysms are not a random sample of those that were present. Instead, cataclysms tend to destroy species that are specialised and localised, leaving widespread, opportunistic generalists intact (Erwin 1998; McKinney and Lockwood 1999). As we noted before, ecological generalists are usually favoured in disturbed environments, while specialists flourish when conditions are stable. If conditions were stable for a long time before the cataclysm, then generalists may have been largely suppressed by competitively superior specialists. Surviving species are also likely to suffer population fragmentation during catastrophes. As we saw above, landscape fragmentation facilitates speciation. In addition, the removal of large numbers of species, particularly ecological specialists, may leave many resources unused. Thus a burst of speciation could follow the catastrophe as isolated pockets of survivors diversify. As the landscape subsequently fills, competition becomes increasingly important for survival. This competition causes species to evolve towards ecological specialization once more. However, many of the new ecological specialists are likely to be descendents not of old specialists, but of generalists who diversified after the last cataclysm.

Experimentally testing theories about the mechanisms behind such large-scale evolutionary patterns remains a challenge. Often, as with the pollen data, arguments rely on inference from indirect historical evidence. Recently, however, some researchers have begun to use computational models to directly test some of their most ambitious theories about the origins and dynamics of complex life and ecosystems. In the next chapter, we shall look at some of these projects in detail.

CHAPTER 9

VIRTUAL WORLDS

A virtual reality model of an agricultural landscape in South East Asia (developed at Monash University by Tom Chandler, Michael Lim, Kenneth Hsieh, and reproduced here with permission).

Some of the biggest questions about life are also the hardest to answer using traditional experimental methods. How did life begin? How did multicellular organisms arise? What are the major transitions in evolution? Why is life so diverse? What might life on other planets be like? These and other big questions deal with processes that occur on massive time and spatial scales. Without a time machine, we cannot directly test our ideas about them empirically. The same problem is faced in other fields of research, from astronomy and geophysics to archaeology and palaeontology.

Likewise, in landscape ecology, there are many experiments that you could never perform. You could not, for instance, cut down a mature forest to see whether it eventually regenerates. Trees grow so slowly that the experiment might take a thousand years to complete, and the experiment might destroy the very system you set out to understand. Such limits on experimentation pose a serious problem for ecologists. Without the ability to experiment, how can we ever know whether environmental management practices will achieve the desired long-term results? One answer is to use virtual experiments: simulation models.

Although we cannot always directly test ideas about ecological processes in the real world, we can at least make sure that our hypotheses are internally consistent. We can examine their underlying logic using computer programs or formulae that explicitly encode this logic. Embodying hypotheses about ecosystems in a model is one of the most severe tests that you can apply. It forces you to pin down your assumptions precisely. Such testing can immediately make clear that some ideas will not work the way we expected. It can also suggest indirect empirical patterns that could falsify our hypotheses. If the model behaves in the way that we predict, then it provides evidence to support our theory. If the model behaves in an unexpected fashion, then it may offer fresh insights into the ecological processes in question.

In this chapter, we will look at when and where models are useful, the hazards involved in applying them, and some of the key insights derived from them. We begin by examining the role of models and simulations in ecology. In previous sections, we have seen many examples of models and the insights that they can provide into processes in landscape ecology. Here, we will extend the discussion to look at some biological models that aim to examine more fundamental questions, not only about ecology, but also about the nature and origins of life. These models arise in the field of artificial life. Lastly, we shall look at virtual worlds as tools for education and exploration in conservation.

9.1 VIRTUAL EXPERIMENTS

Look at a road map. Typically, it will be very simple: lots of dots, representing towns; lines joining the dots to represent roads; and text labels to tell you what objects the dots and lines represent. A road map is a model of the landscape, a simple model, admittedly, and yet a model that serves its purpose admirably. It omits needless detail and conveys exactly the information that a driver needs to navigate from point A to point B. When we come to think about ecological models, it is useful to keep the example of the road map in mind. What is the model meant to do? How much detail is actually needed to achieve that goal? What can be left out without sacrificing essentials?

A *model* is any representation of an object or situation that preserves features of the original. Road maps, photographs, and sculptures are all models in the sense that each represents some aspect of something in the real world. Experimental animals are often used as physiological or behavioural models of humans. Scientific theories are models that are supposed to give true explanations of things that happen in nature.

In general, models have three kinds of uses: *explanation*, *prediction* and *control*. With physical models, this distinction is not important: the one model usually serves for all three applications. However, the complexity of biological phenomena means that we often need separate models for each function.

Explanatory models include any model that tries to help us understand a system. Even the way such models fail can be useful. Statistical hypotheses, for instance, always model some of the properties of variables in a system.

Predictive models attempts to provide estimates for the values of variables when they cannot be measured. Such models need to be accurate, rather than revealing. Perhaps the most common use is forecasting, such as predicting future population trends. Predictive models are often developed by fitting curves to existing data. A crucial principle is to test the performance of such a model using different data (*test data*) from that used to build it (*training data*).

Control models are models that we use in trying to manage a system so that it performs as we would like. For instance, size and catch limits in fisheries are attempts to make a fish population conform to a desired model. Perhaps the most common use of control models is in optimisation, in which managers try to identify conditions under which certain features take on maximum or minimum values. Land use allocation, for instance, is an optimisation problem in which we might try to conserve maximum species diversity in face of various constraints (for example, a required level of forestry production).

One of the most widespread problems of ecological modellers is the failure to state clearly the assumptions that underlie a model. In developing models, it is essential to keep the application (i.e. explanation, prediction, or control) firmly in mind. Models need to be assessed in terms of their adequacy for one of these uses. In developing or assessing models we need to answer the following two questions:

- *Is it valid?* Is the representation real enough, or accurate enough, for the desired goal?

- *Is it useful?* Does it help us to achieve the desired goal? Can we apply it to a problem of concern?

9.1.1 From landscapes to virtual worlds

Imagine that you are playing a video game. In this game, you are walking through a forest. The trees all look real. The scenery changes as you move. You can interact with objects. In every respect, it is easy to believe that you are experiencing a real world inside your computer.

This is the popular image of simulation: the world inside a computer, an artificial world that seems complete in every detail. The intuitive graphical representations that we can create with simulations can greatly help us to understand and communicate the processes we study by giving a gut-level appreciation of the dynamics involved. On the other hand, the feeling of understanding we gain from a visual display may not reflect true reality. There is a great risk of confusing credibility with validity.

One of the authors experienced this danger at first hand when attending a workshop about ecological simulation. His pilot model was scarcely more than a colourful display, but its visual interface allowed the user to see different scenarios being played out. Much to his embarrassment, the audience, which had sat impassively through models that had yielded important results, waxed enthusiastic about his model, chiefly because its real-time display gave it greater credibility than the other, more advanced models.

The questions posed in the previous section are particularly important in simulations of ecosystems. Many people confuse simulation with virtual reality. They think that a simulation model literally reproduces the world, inside a computer. Many early simulation models suffered from this mindset. Their creators literally did try to incorporate every conceivable feature in the models. The great virtue of most of the models we have looked at is that they throw out irrelevant detail and abstract the essence of a natural system. The resulting models allow us to examine key properties of a system.

Another way to look at models is to view them, and especially simulation models, as a form of hypothesis testing. It is all very well to pose a theory that a particular ecosystem does X because of Y. However, if you embody that hypothesis in a simulation model, then you are forced not only to make all your assumptions explicit, but to make them concrete as well. You have to pin down every detail, or the model will not run. So if the behaviour of the model resembles the behaviour of the real system, then in effect it is an experiment in which the result supports your hypothesis.

9.1.2 The need for simulation

As we have seen, models provide one of the most important tools for understanding the consequences of complex interactions. They play a crucial part in the complexity paradigm. We need models because there are many experiments in landscape ecology that we cannot perform in practice. There are several more reasons why we need simulation models as well as mathematical formulae.

A *simulation* is a model that is designed to behave like a system in the real world. Interactive computer games are usually simulations. The aim of simulation is to mimic features of a system that are relevant to particular problems. Simulations never reproduce the real world inside the computer; there are always processes that are ignored, or approximations that have to be made.

A computer simulation is a precise and formal way of representing ideas about how a system works. It has the desirable effect of requiring us to state hypotheses and assumptions explicitly. This enforces clear thinking about what variables and processes are truly important to the problem at hand. Simulation also makes possible explicit 'experiments' to test hypotheses embodied in the model, especially useful where real-world experiments are impossible. For example, in a few minutes a computer simulation can test the severity of a nuclear winter that might follow a nuclear war, or carry out an experiment on evolution that might take millions of years in nature.

The greatest difference between a computer simulation and other models (for example, models that are expressed as mathematical formulae) is that simulations mimic processes, rather than relationships between variables. In models expressed as formulae, mathematical analysis often reveals insights about the nature of the process, but it is not always practical to describe a real system solely in terms of the relationships between variables. In contrast, simulation models are highly flexible, and can be used even where an algebraic representation is impractical or intractable.

Complex formulae are often intractable. The equations cannot be solved. The only way to generate solutions is to express the formulae in simulations and calculate values step by step. Moreover, ecological systems are often inherently unpredictable, so it is not sensible to use formulae to predict outcomes. In such cases, simulations enable us to ask 'what-if' questions about many possible scenarios. Lastly, in many situations it is important to be able to interact with a system to get some sense of what outcomes may stem from different actions.

A good example is the behaviour of wildfires, which are notoriously difficult to predict. For a start, ignitions depend on many, varied phenomena (especially lightning strikes and campfires getting out of control) that make it impossible to foresee exactly when and where they will start. Likewise the pattern of spread depends on many imponderables, such as wind velocity, fuel distribution, temperature, humidity, and fuel moisture.

Two ideas closely associated with the testing of simulation models are sensitivity analysis and scenarios. *Sensitivity analysis* is the process of systematically testing how changes to assumptions and parameter values affect a model's outcome or performance. Such tests are necessary both to check a model's validity and to discover what makes the modelled system behave the way it does. Also, because complex systems often display chaotic behaviour, it is important to detect whether they are sensitive to initial conditions. Because simulation models often deal with complex systems that are inherently unpredictable, we use them instead to examine what might happen under particular conditions. Tests of this kind are known as *scenarios*.

9.1.3 A world inside a computer?

As we have said, most people, influenced by popular exposure to flight simulators and other video games, and by movies based around the theme of simulation, think of simulation as recreating the real world inside a computer. However, to believe that simulations must always be like this is a misconception. Simulation models are often wrongly judged by what they leave out, rather than what they keep in.

A simulation is a model, an abstraction. Like a road map, it is reality stripped of needless detail to focus on the essential elements and processes. Most of the cellular automaton models presented in this book are of this kind. They strip away much of the detail so that the effects that emerge from interactions can be seen all the more clearly. For instance, a model designed to explain the effects of seed dispersal need not include every ecological process, but should start from the simplest set of assumptions until the need for complex explanations is demonstrated.

Once you have a simulation model, you can study the properties of that abstract system and then see what the results tell you about the real world. Unlike the real world, you can perform virtual experiments at will. In this way, you can test out scenarios that would be potentially disastrous, or too expensive, or too slow, to try in the real world. You can vary factors systematically to see what effect they have, and you can repeat an experiment thousands of times to discover the relative frequency of different outcomes. For example, as we saw in Chapter 3, with a model of fire spread, you can test out scenarios, such as the effectiveness of fire breaks or of fuel reduction burning. You can also run hundreds, or even thousands of virtual fires to assess what effect (say) temperature, or rainfall, or wind speed has on fire danger.

An important, but often overlooked, application of simulation is to provide guidelines for interpreting data. An example from traditional ecology will serve to show how this works. Statistical models provide perhaps the most common instances of using models to help interpret ecological data. For instance, in the absence of any other factors, you might expect that trees would be randomly distributed in a landscape. However, if successive generations of trees grow up from seedlings around their parent trees, then you would expect to find some clustering. Comparing the results of sampling from models of the two patterns can give you an indication of how to detect the differences in real data. We have seen some examples of the use of simulations in this book. In Chapter 7, for instance, we saw how simulations were used to test hypotheses about the processes involved in the formation of stable food webs.

9.1.4 Just So Stories?

In the 1800s, the English author Rudyard Kipling wrote a series of tales called *Just So Stories*. These were tales designed to illustrate the rightness of various moral principles. Simulation has sometimes been compared to 'just so stories'. As we saw above, the purpose of modelling is to test our ideas about how a system may work. Unlike empirical researchers, modellers have absolute control over their experimental system. There is therefore always the danger (whether intentional or not) of building a system that, rather than challenging our ideas, simply appears to confirm them. Like the author of the *Just So Stories*, we can always construct a model that appears to demonstrate our convictions, but that does not mean that our convictions are correct.

This issue reflects a fundamental question: how can we be sure that models truly represent what is going on in real ecosystems? This is the validation problem. Modellers usually resort to simulation when a process or system is so complex that analytic models are intractable anyway. Given that a complex system may be unpredictable, no model is ever going to act precisely like the real thing; so even a valid model may not be capable of prediction. Validation can take several forms. One method is to validate the mechanisms built into a model. For example, we can experimentally determine that the neighbourhood function of a cellular automaton accurately represents the dispersal pattern of a certain species. Another method is to validate the general adequacy of the model for the use for which it is intended. Ecological models may be suitable for either prediction or explanation, but not necessarily both at once.

The issue of scale has a great bearing on the way a model is represented, and sometimes more than one model may be needed to understand the same process at different scales. For example, in developing possible control scenarios for the introduction of foot and mouth disease into Australia, two models were required (Pech et al. 1992). One considered the large scale picture, with probabilities of spread between cells occupied by feral pigs. A finer scale model, which tested assumptions built into the large-scale model, incorporated individual pigs within an environment and dealt with the problem of rare events: the transmission of disease from one individual to another.

Similarly, simulations of bushfire spread are virtually useless for fire fighting. Studies of fire behaviour, including calibrating and validating models of spatial spread against experimental burns, show clearly that the pattern of spread in a patchy fuel bed is highly unpredictable. Fire behaviour is also highly sensitive to variations in parameters such as wind direction and fuel moisture, both of which can change rapidly as a fire burns. In any real fire, these factors are never known in sufficient detail to make accurate prediction possible. On the other hand, such models are excellent tools for planning and training. They allow researchers to perform experiments and test scenarios that would be impossible in reality.

9.2 WHAT IS ARTIFICIAL LIFE?

Have you ever seen one of those old movies about the Stone Age in which prehistoric people are pursued by dinosaurs? The immediate objection is that humans and dinosaurs never coexisted. Dinosaurs became exist almost 65 million years before humans evolved. Nevertheless it is fascinating (as popular movies prove) to wonder how people would cope if confronted by a Tyrannosaurus.

In the same way, we can ask what would happen if other combinations of plants and animals were thrown together. Asking these kinds of what-if questions belongs to the realm of simulation. However, why restrict ourselves to questions about real plants and animals? Instead of playing with life as it is, we can play with life as it could be. Why be satisfied with mere dinosaurs, when we could have Godzilla running around?

There is a serious side to such speculation. Carrying out virtual experiments with worlds that do not exist may teach us lessons about why the real world, and real living things, are the way they are. Could any ecosystem actually support a

thousand tonne carnivore? Answering such questions, by experimenting with life
as it could be, is the province of the new field of Artificial Life.

Frustrated by the fragmented nature of the literature on biological modelling
and simulation, an American scientist, Chris Langton, decided to try to bring all
the researchers together. He coined the term *Artificial Life* (Alife for short) and
organised the first Alife conference, which was held at Santa Fe in 1987. The
meeting proved to be a great success. Perhaps its most significant aspect was that
scientists from many different disciplines found that they were working on related
problems in total ignorance of what was going on in other areas. Alife conferences
have been held biennially ever since, and the field has grown into a very active
area of research. An international Alife Society was formed in the year 2000.

Artificial Life (Alife) tries to understand living systems by imitating them in
simulation and other models. It not only addresses life as it is, but also *life as it
could be*. Experimenting with virtual worlds allows us to explore all the
possibilities of nature, and thereby to better understand the world as it is.

Typically, Alife models deal with multi-agent systems. As the name implies, a
multi-agent system consists of many independent agents that can interact with
each other, and with their environment. In computing parlance, the term *agent*
refers to any piece of software that acts independently, without direct control.
Agents can also interact with other agents. In the context of ecological simulation,
an agent would usually represent (say) an animal such as a lion or a zebra moving
around in the landscape, or a human who interacts with the ecosystem in some
way. In cellular automata models, we can regard each cell as a distinct agent that
interacts with its neighbours.

In terms of its impact on theory, arguably the greatest achievements of Alife
lie in its revelation of the ease with which self-organisation can occur under
certain conditions. The patterns that emerged in early Alife models suggest that it
may be possible to explain the complexity of many kinds of biological
organisation in terms of simple rules. The challenge is to translate that potential
into models of real systems. In earlier chapters we have already seen quite a few
applications of Alife models. Examples include the models of boid flocks, bee
societies, plant growth and starfish outbreaks considered in Chapter 2, the cellular
automata models of landscapes discussed in Chapter 3, and in Chapters 6 to 8, and
the agent-based models of speciation in Chapter 8. In the following sections we
will look at some classic Alife models that have not been introduced previously.

Perhaps the most important practical effect of ALife has been to promote the
paradigm of natural computation. Computational models of biological issues find
much greater acceptance today than 20 years ago. At the same time, some aspects
of advanced computing have become almost indistinguishable from biology.
Simplistic though they often are from a biological standpoint, ALife models
provide insights into the ways in which living systems solve complex
computational problems. We saw an example of this in the ant sort algorithm in
Chapter 2.

In the early days of Alife research, the questions addressed were very different
from the concerns of mainstream ecology. More recently, however, Alife
researchers have begun to attack some of the classic questions in ecology,

providing novel angles on issues, such as group selection, that have been a source of hot debate amongst biologists (see review by Wilke and Adami 2002). The field has been dominated by a number of seminal models. Each of these models addresses fundamental, unresolved issues in theoretical biology. One of the earliest Alife models was the *Game of Life*, which first drew widespread attention to the possibility of simulating life-like processes using simple computational methods and exposed a number of universal features of multi-agent systems. The *hypercycle* theory (Eigen and Schuster 1979, Boerlijst and Hogeweg 1991) suggested mechanisms by which early chemical systems began the transition to self-replication. The *N-K* model provided perhaps the first insight into the ways in which genes might be organised to exercise control over development and metabolism (Kauffmann 1993). The *Tierra* model highlights some of the issues in early evolutionary systems, such as the appearance of viruses, the robustness of key ecological patterns, and the possibility of spontaneous extinctions. Lastly, the *Daisyworld* model has been a key tool in developing the Gaia hypothesis and the view of ecosystems as complex adaptive systems. We will look in more detail at some of these models below.

9.2.1 The Game of Life

One of the first, and perhaps the most famous Alife model is the Game of Life, devised by the Cambridge mathematician John Conway (Gardner 1970). The game demonstrates several general features that are common in systems of many agents. The most sweeping of these is that simple rules impose order (Figure 9-1), but at the same time are capable of producing very rich behaviour (Figure 9-2).

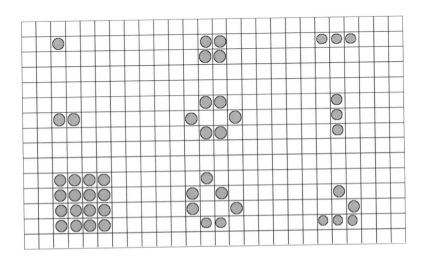

Figure 9-1. Some of the complex self-replicating forms that can be generated in the Game of Life.

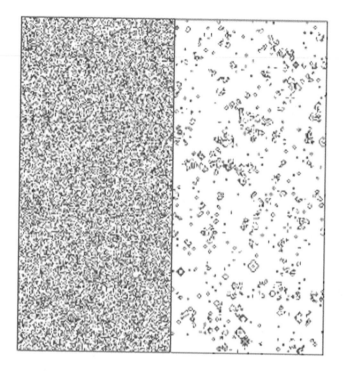

Figure 9-2. The Game of Life. The left panel shows the chaotic initial state of the system, while the right panel shows its state after many generations. Note the complex, but non-random patterns that have been generated over time.

The game is a cellular automaton model in which each cell exists in one of two states: "alive" or "dead". Likewise, the rules of behaviour mimic various life-like properties. The neighbourhood of each cell consists of the eight cells in the 3×3 box immediately around it. The model goes through a sequence of generations in which the states of the cell progressively change. The rules governing the transition from one generation to the next are simple:

1. If a living cell has less than two living neighbours, then it dies from loneliness;

2. If a living cell has more than three living neighbours, then it dies from overcrowding;

3. If a dead cell has exactly three living neighbours, then a birth occurs.

The Game of Life proved extremely popular for many years, especially amongst computer programmers, and many versions are widely available. It was the first program to demonstrate what were later to be recognised as common features of cellular automata, and of complex systems in general. The most telling of these features relate to the final state of the system. In a finite universe, any random starting configuration of live cells eventually gives rise to just two kinds of end results (both represented in the right-hand panel of Figure 9-2):

1. static patterns that never change;

2. cyclic patterns that repeat at a constant interval.

These patterns are attractors in the model, just like the attractors described in Chapter 3. The early patterns, and most of the interest in the game, centre on the transient behaviour that it undergoes before settling into these attractor states.

The above observations about the Game of Life led to research into the computational properties of many kinds of cellular automata, most notably by Stephen Wolfram (1986) and later by Christopher Langton (1990). The studies revealed four distinct classes of behaviour: fixed, periodic, chaotic and complex. These behavioural attractors are related to equilibria and limit cycles in dynamic systems. Any finite cellular automaton must ultimately fall into a fixed state or cycle. This is because a cellular automaton consisting of N cells, with a possible S states each, has a total of N^S possible configurations of cell states overall. So it is inevitable that the cellular automaton must eventually repeat some earlier configuration after at most N^S generations.

The behavioural richness of a cellular automaton is related to the number of neighbourhood configurations that force a state change in the cell at the centre of the neighbourhood. In the Game of Life, for instance, there are 512 possible configurations for a cell and its eight neighbours. Of these, the game's rules force the state of the cell to change in 216 cases. In general, models that force a change of state for very few configurations tend to freeze into fixed patterns, whereas models that change the cell's state in most configurations tend to behave in a more active 'gaseous' way. That is, fixed patterns never emerge. Systems with intermediate levels of activity usually exhibit the most interesting behaviour. For instance, it has been shown that the Game of Life is capable of universal computation.

The Game of Life is the prototype cellular automaton. It has served as the inspiration for the investigation of many other cellular automaton systems and for many of the kinds of applications that we looked at in earlier chapters, where we saw cellular automaton models of fire spread, epidemics, forest succession. All of these are Alife models.

9.2.2 Tierra

Another challenge for biological theory was to understand how a self-replicating system could exist at all. How is it possible for a system to include all the details needed to reproduce itself? Surely the information required amounted to a complete description of the individual, including the description of itself? Like a pair of mirrors reflecting each other, the problem seemed to spiral off into infinity. And yet, somehow, living organisms clearly do resolve this paradox.

The solution seems to be that organisms do not need to encode a complete description of themselves; they just need the recipe for making a copy (i.e. an embryo that grows into a complete organism). Most of the detail is supplied by the context (i.e. during development). Early investigators into self-replication, notably John von Neumann and Lionel Penrose, demonstrated that relatively simple systems could replicate themselves without the need for complex biological machinery.

In the light of intense interest in how self-replication first arose in living systems, an important step was to demonstrate that self-replication could occur in simulated organisms. Tom Ray's model Tierra did exactly this (Ray, 1991). Tierra consists of 'organisms' that are programs which include code allowing them to replicate themselves. Their environment is a computer memory, and as in a real environment, space is limited. The organisms therefore have to compete for space within this finite universe.

One of the most notable findings of the Tierra model, and the inspiration for much later research, was the revelation that organisms and ecologies of increasing complexity did not emerge automatically from the evolving system. Instead, competition drove the digital organisms to become as small as possible, and ecological interactions remained very limited. Later researchers were able to develop Tierra-like models where the complexity of individual organisms did increase, but only by freeing the digital organisms from any realistic energy constraints and explicitly selecting for complexity (Adami 1994). These findings suggest that Tierra-like models lack some fundamental properties of real biological systems. Simultaneously, the Tierra models did demonstrate that some features of ecological and evolutionary systems are highly robust: for example, realistic relative abundance distributions, varieties-area relationships, and possible self-organised criticality were observed (Adami and Brown 1995, Adami et al. 1995).

9.2.3 Daisyworld

In 1974, James Lovelock and Lyn Margulis proposed a startling and controversial theory. Their *Gaia Hypothesis* proposed that the planet Earth acted as a kind of super-organism, which regulated the environment to keep conditions suitable for life to flourish (Lovelock 1989). Not surprisingly, many scientists were put off by the teleological approach implied by the notion of a super-organism, and by the religious overtones suggested by the name Gaia (the Earth Mother goddess). Nevertheless, the theory did suggest an important hypothesis about stability of global climate (Lovelock and Margulis 1974).

Essentially, the theory boils down to a matter of negative feedback. What the theory says is that organic, geologic and atmospheric processes combine to dampen changes in conditions within the biosphere. Take oxygen, for instance. Oxygen accounts for about 21 per cent of the earth's atmosphere. If you increased the level just a few percentage points, then vast forest fires would rage and consume the difference. If you decreased the level just a few points, then many creatures would start to die from oxygen starvation and plants would soon restore the balance.

The most damaging criticism of the Gaia hypothesis has been the initial absence of any clear mechanism. Biological organisms evolve homeostatic mechanisms through natural selection. As we saw in Chapter 8, natural selection acts on systems that consist of units which reproduce themselves with variation.

The Earth, however, does not reproduce and therefore it is unclear why it should evolve to a homeostatic state.

The Daisyworld model was Lovelock's first attempt to formally address this criticism by illustrating how simple biogeographical processes might contribute to the regulation of global temperatures. The model starts with a hypothetical planet whose surface reflects back into space a certain fraction of incoming light (its 'albedo') from its sun. For simplicity, the albedo is assumed to be constant everywhere. Growing on the planet are two types of daisies. One species ('white daisies') has an albedo greater than the planet's surface. The other species ('black daisies') has an albedo lower than the planet's surface.

Consider what would happen if, over a period of time, the sun's luminosity changed. If the planet were barren, the surface temperature would steadily change with the sun's strength. However, provided both black and white daisies are present, the dynamics of the daisy populations act to dampen any changes in temperature. When temperatures are low, the sunlight is too weak for the white daisies, which in most areas are out-competed by the black daisies and are consequently confined to a narrow band at the equator. Because black daisies cover most of the planet, the albedo of its surface decreases. So the surface absorbs more sunlight, thus raising the average global temperature. As the sunlight increases, the white daisies become more competitive. Their distribution gradually expands and the black daisies become confined to bands around the poles. The shifting balance in daisy populations also increases the overall surface albedo, thus keeping the surface temperature constant.

Eventually, of course, the balance breaks down if the sunlight increases too far. First the black daisies are driven to extinction. When the white daisies are no longer able to withstand the powerful sunlight at the equator, a runaway effect begins. The surface albedo starts to drop, so the temperature finally starts to rise. Soon, the white daisies are confined to the poles, and eventually disappear, after which the temperature continues to rise in proportion to the sunlight. From this time on, the planet is lifeless.

Cellular automata were used to explore the dynamics of the Daisyworld model in detail, revealing that such systems often maintain homeostasis well under normal conditions but are prone to sudden collapse in response to gradual forcing (Ackland et al. 2003).

Critics remain unconvinced by Daisyworld because it is, they argue, merely one possible system among many (Kirchner 2002). Why should Earth behave like Daisyworld? Several recent models suggest that natural selection can give rise to Gaia-like systems (for example, Lenton and Lovelock 2000, Sugimoto 2002, Staley 2002, Ackland 2004). In addition, there is increasing attention to the parallels between Gaia models and *complex adaptive systems* (sometimes referred to as CASs) (Lenton and van Oijen 2002).

The complex adaptive systems concept is an attempt to abstract the properties of systems that adapt through a form of natural selection. Biological organisms are the most obvious example of these systems. The definition also encompasses a number of other systems, such as neural connections in brains, which are reinforced when used, and die back when unused. A number of authors argue that ecosystems are complex adaptive systems. According to this view, ecological interactions that involve stabilizing feedback processes are likely to last longer and spread further than their competitors. For example, some ecosystems might contain feedback loops which, when confronted with a perturbation, tend to restore the original state of the system. These ecosystems will then survive the perturbation with minimal change. Other ecosystems might be changed drastically by the same perturbation. If such variation exists in ecosystems, then over time we should expect more stable ecosystems to evolve simply because less stable ones change more often. Over long periods, we might expect each local area's ecology to adapt to restore conditions favourable to itself in the face of normal perturbations (Staley 2002). In support of this idea, a number of apparently homeostatic ecological mechanisms have now been identified in global climate systems (Kleidon 2002). We will look at these processes in more detail in Chapter 11 when we discuss the prospect of human-induced global climate change.

9.3 FROM VIRTUAL TO REAL

Models such as we have seen above may help to explain the kinds of processes that occur in ecosystems, but when you try to model a real system, you immediately encounter severe practical problems. Perhaps the greatest of these is how to integrate the simulation with real data.

Putting a simulation model together can be a long and expensive job. First and foremost, there is the problem of writing a computer program. Most ecologists are not skilled at writing computer programs, so this stage immediately adds a large degree of time and cost, since the ecologist needs to find (or hire) a friendly programmer. A more subtle problem is that the ecologist has to communicate his or her intentions clearly to the programmer. Otherwise, the resulting model may not embody what was really intended.

Not surprisingly, the need to simplify the business of building simulation models has led to the appearance of many modelling systems. Many of these are specialised packages that deal only with particular kinds of systems or issues. Typically, simulation packages provide tools to help build models, but do not allow for integrating them with real data resources. Conversely, geographic information systems (GIS) and other databases usually offer only rudimentary modelling capability. A typical GIS model just expresses data in a set of layers. However, several flexible modelling packages that can be integrated with GIS have been developed and are discussed in detail by Fall and Fall (2001). Chapter 10 provides further details about GIS models and other ways to handle large-scale ecological data. In the sections below, we shall look briefly at some examples of virtual reality and agent-based modelling systems.

From the user's point of view, modelling and simulation packages need to provide several "user-friendly" features. First, they should provide convenient ways of defining and implementing models, such as menus or ecologically oriented scripting languages. Secondly, they need to address the problem of integrating models with geographic and other real-world data. Finally they should provide a visual interface that displays model outputs in real time. At the time of writing, few existing systems provide all of these features.

9.3.1 Swarm

One of the practical problems faced by early Alife modellers was the lack of tools for developing simulation models, especially ones involving systems of many agents. To address this issue a workshop was held in the mid 1990s. The outcome was the shell for a software platform for modelling multi-agent systems. The software, called SWARM, provides many tools to simplify the development of models, especially the visual display of outputs, which is always the most time-consuming stage of model development (Langton et al. 1997). Alife researchers have developed a wide range of applications in SWARM. These include economic models, such as Sugarscape, in which agents compete for a limited set of resources, as well as ecological and evolutionary simulations.

9.3.2 Smart Forest

One industry that has been placed under a lot of pressure by increasing environmental concerns is forestry. A positive response to this pressure is the program SmartForest (UIUC 1998). The idea behind the program is to provide a sound basis for assessing the claims about environmental impact made by forestry companies and other groups. At present, a forestry company may as argue: "We do have to cut down part of the forest, but it will not spoil the aesthetics of the area and the patch will recover within twenty years." The argument has much more force if people can actually see what the area will look like. This is exactly what Smartforest is intended to do.

The program combines GIS, simple forest models, and virtual reality graphics. Once a model has been set up, the user can move around in the environment and look at the view from any place, in any direction, at any height from ground level to hundreds of metres. Several other commercial packages of a similar nature are now available.

9.3.3 Ecos

Ecos is an artificial life modelling package developed by Russell Standish at the University of New South Wales. It simulates multi-species interactions in a landscape. Imagine a set of populations that co-exist within an ecosystem. At the population level, we can represent the dynamics of such a system by three sets of parameters. The first group (*alpha*) are the growth rates of the individual populations. There may also be a group of limits (*lambda*) on the carrying capacity of the environment to support each population. A second group of parameters (*beta*) are the interactions of the populations with each other (in a set of N species, there will be N^2-N of these parameters). The third group of parameters (*gamma*) are the diffusion coefficients. In a uniform environment, there would be N such parameters (one per species), but in a heterogeneous landscape they can vary from place to place.

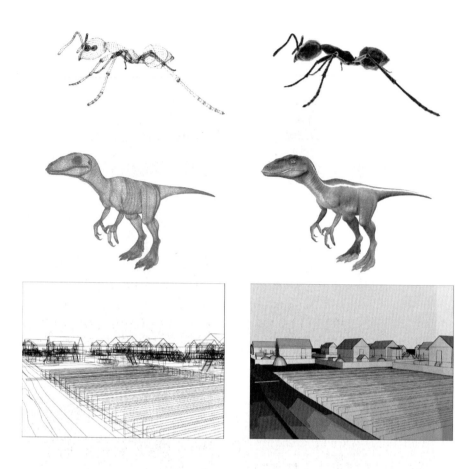

Figure 9-4. Use of wire frames in constructing virtual reality models of an Argentine ant (see photograph at start of Chapter 8), a raptor, and the Southeast Asian agrarian landscape shown in the previous figure. These images were developed at Monash University by Tom Chandler, together with Anders Messel, Knut Eivind and Matthew Rowe (ants), Chandara Ung (raptor), and Michael Lim, and Kenneth Hsieh (landscape). Reproduced here with permission.

If internet uses can explore alternate universes online, and interact with other explorers in the process, then why not develop virtual reproductions of the real thing? Many areas, such as city centres, are already mapped in great detail. By this we mean that the exact forms of the buildings are well-known. Tourist maps of (say) downtown London sometimes include pictures of the buildings. Why not turn these picture maps into online virtual worlds? Tourists or students, for instance, could explore an area to get the feel of a place before they even set foot there, or explore remote areas, such as the Rocky Mountains, in greater depth than they ever could in real life.

CHAPTER 10

DIGITAL ECOLOGY

Rainforest near Kangaroo Valley, southern New South Wales. Rainforests and coral reefs are among the ecosystems that are richest in biodiversity.

10.1 INFORMATION AND COMPLEXITY

Late in the 1990s, Nick Klomp was studying short-tailed shearwaters, one of the most common bird species along Australia's eastern coastline. During the breeding season, the shearwaters build their nests on off-shore islands. While one parent tends the nest, the other flies out to sea in search of food. A crucial question in Klomp's research was how the bird populations interacted with the marine species they feed on. He used a combination of tracking satellites and small transponders attached to the birds to track where they flew within their nesting territories. To everyone's surprise, data returned by the new technology showed that instead of making short sorties close to their nests, the shearwaters undertake long migrations (Klomp et al. 1997; Klomp and Schultz 2000). They travel several thousand kilometres to and from Antarctica each year during the southern hemisphere summer (Figure 10-1). Klomp's discovery had many implications, not only about the ecology of shearwaters, but also about species interactions and food webs in Antarctic waters.

This story is a good example of unexpected discovery or serendipity. Serendipity is extremely common in science and is an almost inevitable result of approaching scientific problems in new ways (Green 2004). The word 'new' can mean many different things here. It can, for instance, mean applying new technology, investigating an ecosystem for the first time, or simply asking new kinds of questions.

Radio tracking technology has led to many surprising discoveries. Great white sharks, for instance, migrate all the way from California to Hawaii (Boustany et al. 2002). The same technology has been used to track movements of many other species, including dolphins (Würsig et al. 1991), polar bears (Armstrup and Durner 1995), and manatees in the Florida Everglades (Weigle et al. 2001; Deutsch et al. 2003).

As we have seen in earlier chapters, applying new ideas regarding complexity in ecology leads to new insights about landscapes and ecosystems. However, this new approach to landscape ecology is possible only because new technology, namely computers and information systems, have made it possible. Complex systems science is in large part a result of the information revolution. The emerging new complexity paradigm that we mentioned in Chapter 1 is intimately bound up with advances in information technology. To deal with ecological complexity in practice, land managers need information. They need it in abundance. They also need the capability to distribute, store, manipulate, and interpret ecological and geographic information.

The new ideas and methods arising from advances in information technology themselves constitute a new paradigm, which goes under various names such as informatics, or e-science. The information revolution has sparked enormous changes in the ways scientists do their work. The biggest change is the ability to handle data. In times gone by, data were rare and expensive. Today they are increasingly abundant and cheap. That transition has wrought major alterations in the way science is done. It also makes possible practical management of ecological complexity.

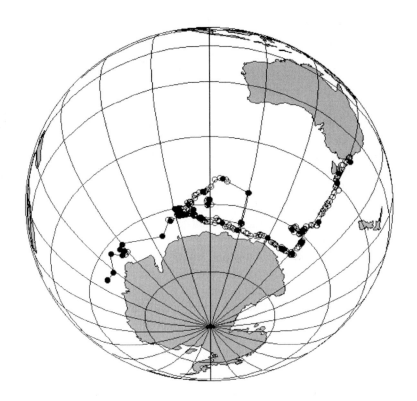

Figure 10-1. Satellite tracking data revealed that short-tailed shearwaters, during breeding on Montague Island in southern New South Wales, undertake epic foraging journeys across the Great Southern Ocean (Klomp et al. 1997; Klomp and Schultz 2000).

In this Chapter, we explore some of the ways in which the information revolution is transforming ecology, especially its role in the study of complexity in landscape ecology.

10.2 THE COMPLEXITY PARADIGM

"The complexity revolution is aptly named, ... for it changes the way that scientists approach their subject matter. It has given us new tools and concepts, as well as fresh explanations for age-old puzzles" (Davies 1998).

Early in the 20th century ecologists became more and more interested in the notion of 'diversity'[1]. Some ecosystems, especially rainforests and coral reefs, seemed to be teeming with a great variety of species. Others, such as the Arctic

---

[1] For a detailed account of diversity and the way it is measured, see Pielou (1975).

wastes, seemed to have very few inhabitants. As ecology became increasingly quantitative, ecologists needed ways to make this idea of diversity more precise. They soon devised ways to quantify the notions of species richness and variety. The first and most obvious measure was the number of species, which we can think of as measuring biological *richness*. Taken on a large scale, the number of species continues to be the most widely used indicator of biodiversity. At the ecosystem level, the idea of species richness does not account for everything: some communities have a more varied mixture of species than others in which one or two species are super abundant. Ecologists therefore found it necessary to introduce other measures to express different interpretations of diversity, such as uneven population sizes and habitat variety.

In landscape ecology, as in every area of science, there is a close relationship between theory and data. The theories that scientists develop are limited by the data available to them. Conversely, the data that scientists collect are determined by the theories that they wish to test and by the experiments, or observations that they carry out. We can look at the concept of diversity to see how this interaction between theory and experiment happens.

By expressing such a fundamental idea in quantitative terms, ecologists had a tool with which they could state ecological hypotheses in precise fashion. This ability in turn made it possible to devise precise tests of those hypotheses. It also made clear what kinds of data they needed to collect.

As a research tool, diversity indices can be useful in revealing biological relationships. They have also helped ecologists to study connections between a community's diversity, its environment, and its internal composition and processes. To take just one example, Recher (1969) in Australia and Cody (1970) in Chile showed that bird species diversities were significantly correlated to foliage height diversities, implying that birds chose habitats horizontally in different vertical layers of the forest as well as in different localities.

Having been formally defined, diversity has come to be seen as a fundamental property of communities that can be studied in its own right. Like mass or energy in physics, it is an abstract concept that can be discussed independently of particular communities, allowing us to uncover general rules of ecology. Thus, for example, we know that diversity normally increases during succession, that communities in harsh environments (for example, mountain tops) are usually less diverse than comparable ones in more benign surroundings. Diversity studies have also provided evidence bearing on controversial areas of ecological theory, such as island biogeography and the relationship between stability and complexity in ecosystems.

As we saw in earlier Chapters, people have often used diversity as an indicator of ecological complexity. The reasoning is that high species diversity implies a greater richness of interactions between species, which in turn leads to more complex behaviour. Likewise, high habitat diversity implies a rich and complex range of interactions between sites. The story of ecology's attempts to deal with diversity therefore reflects, in part, its attempts to come to grips with complexity. As we saw in Chapter 7, the diversity-stability issue forced ecologists to re-examine many traditional assumptions. It also forced scientists to focus on the nature of complexity by looking in particular at the underlying structures of food webs.

Several other measures or indicators of complexity can be used. As we have seen, another possible measure is connectivity. Earlier we saw how this applied to landscapes (Chapter 3) and to food webs (Chapter 7). Other measures, such as clustering coefficient and path length (Chapter 7) reflect structural properties of complex networks.

10.2.1 Scientific paradigms

Scientific research does not happen in a vacuum. The way in which people look at the world conditions the questions that scientists ask, the experiments they perform, the data they collect, the way they interpret their results, and ultimately, the kinds of theories they develop. The prevalence of such influences led Thomas Kuhn and others to argue that science is dominated by paradigms (Kuhn 1962; Popper 1968). Paradigms are bodies of theory and methods that together define a particular way of looking at the world.

In Chapter 1, we argued that the new ideas and methods that have been developed to deal with complexity in landscape ecology amount to a new paradigm. In subsequent chapters, we have described many ideas and methods associated with this paradigm in the course of discussing the effects of complexity on landscapes and ecosystems. This is not to imply that current paradigms need to be replaced or forgotten, but rather that they need to be complemented by different approaches to deal with different issues. In particular, complexity theory recognises that:

- interactions lead to richness and variety in ecosystems and landscapes;
- global phenomena often emerge out of local interactions and processes;
- many processes are based on qualitative effects that need to be captured by symbolic rules, rather than by numeric formulae;
- no single model of a complex system can be all-embracing, so different models may be needed for different purposes, whether it is to explain, to predict, or to control.

However, we have not yet sketched out how the complexity paradigm works in practice. By this stage, for instance, any field ecologist will be asking how they can come to grips with ecological complexity. What questions do you ask? What experiments can you do? What data should you collect? And once you get that data, how do you go about interpreting it?

Some of the differences between the complex systems paradigm and traditional approaches are evident if we compare simulation with traditional approaches. Just like a mathematical equation, simulation is a tool that scientists can use to represent patterns and processes in nature. The difference is that equations are solved, whereas simulations are played out. Because complex systems are often inherently unpredictable, we examine scenarios instead of making forecasts. Instead of solving equations, we perform sensitivity analysis.

Instead of using analytic methods to find optimal solutions, we use genetic algorithms (Holland 1975) and other adaptive methods. Instead of plotting graphs, we use visualisation or virtual reality. Finally, because relationships abound in complex systems, simple relationships may not be evident, so exploratory methods, such as data mining, are increasingly important in research.

Many common kinds of questions associated with complexity involve patterns of connections. As we saw in Chapter 7, for instance, food web studies go beyond simplified or aggregated classifications, such as trophic levels[2], or predator-prey interactions, and look in detail at the entire pattern of interactions within an ecosystem. Likewise, many studies of landscape habitats and species distributions now include analysis of spatial connectivity and its effects.

10.3 FROM FIELD WORK TO ECOTECHNOLOGY

In 1963, the American ecologist Joe Connell began an unusual study of two areas of Queensland rain forest, at Davies Creek and O'Reillys[3]. Over patches of one to two hectares, he and his colleagues Geoff Tracey and Len Webb mapped the location of all mature trees, noting carefully the size, species, and other features of each one, as well as mapping, tagging and measuring every small tree and seedling. This survey involved recording thousands of individual trees and seedlings and required weeks of effort to complete. Their project did not stop there. Every few years, they came back and did it all over again. In time, many other colleagues joined in the work. Eventually, in 1993, the monitoring study became a permanent project of the CSIRO's Tropical Rain Forest Centre. At the time of writing, the project has traced the fate of tens of thousands of individual trees and seedlings over a forty year period. In doing so, it has created a unique record of spatial and temporal change within a patch of rainforest, which enabled the ecologists to observe many processes, such as seedling recruitment, and to test hypotheses, such as compensatory recruitment.

Although there are other monitoring projects on a similar scale around the world, few ecologists can afford the huge amount of time and effort required to map even a small patch of forest just once in such detail. Most ecologists have to make do with samples of the landscape. Traditional methods of data collection in ecology are based on the need to glean maximum information from limited data. Often the field methods involve a compromise, originally devised because a great deal of effort was needed to win every scrap of raw data.

One of the principal field methods in traditional ecology is the *quadrat* (see Chapter 3, Section 2). This is a small area within which the ecologist records

---

[2] In traditional ecology, populations that comprise an ecosystem fit into a trophic pyramid, with producers (e.g. plants) on the bottom layer) and top predators at the top.

[3] A detailed account of these rainforest monitoring projects can be found online:
http://www.lifesci.ucsb.edu/eemb/faculty/connell/research.html

every relevant detail[4]. To gain a representative picture of the landscape, field ecologists combine the results that they obtain from a number of randomly distributed quadrats. To take account of systematic variations across a landscape, such as changes in elevation, they string a series of quadrats into *transects*. These methods form part of a traditional paradigm that is based on statistical sampling and hypothesis testing.

There are also traditional field methods to take interactions into account. For instance, neighbouring trees interact with one another. Most obviously, their canopies compete for light. Below ground, their roots might compete for water. And then there are less obvious kinds of interactions, such as the competition between their offspring for open ground on which to grow.

Given these, and many other, possible interactions, it is important to know something about a tree's neighbours. But in quadrat sampling, a tree's neighbours are often outside the sampling area. So ecologists traditionally approximate. They reason that the closest neighbour is the one likely to have the greatest effect. As a way of investigating interactions within a landscape, nearest neighbour records provide at best a rough approximation of what is going on. For instance, a large tree further away might create more shade than the nearest neighbour. Although there are refinements to cope with such problems, such as recording several near neighbours, they are always a compromise. On the other hand, if you compile a complete map of all trees in an area, then not only can you compile nearest neighbour data by reading the map, but also you can produce any other indicators you wish, even if they do not occur to you until later.

Similarly, ecologists usually go about collecting data by making a field trip to the area of interest. However, a single field trip to a particular area is just a snapshot that captures conditions at a single instant in time. To understand the processes that really drive an ecosystem, there is no substitute for continued monitoring over a long period of time. Few researchers can afford to devote the time and resources needed to emulate Joe Connell and his colleagues by monitoring an ecosystem, or even part of one, over many years. Field trips, especially to remote areas, can be expensive and require a considerably investment of time and effort. Most people have to make do with a handful of field trips over a limited time period.

As a consequence, ecologists often feel constrained to answer only relatively simple questions about their data. For instance, has the system changed? What changes occur together? Observing processes in detail over time, especially long periods of time, is often not possible. In other words, the limitations imposed by all the difficulties surrounding traditional field ecology make it impossible to address many important ecological questions. In particular, it has been impossible for ecologists to come to grips with the complex patterns and processes in ecosystems that we have discussed in this book.

---

[4] The size of the quadrat varies enormously. Ecologists studying (say) small plants, seedlings and soil organisms might use a square frame, one metre by one metre in size, whereas in surveys of large animals, the quadrats are sometimes as large as a square kilometre.

Over the years, however, the situation has changed. For one thing, new technologies have made it possible to gather more data, more quickly than ever before. Also, the gradual accumulation of data makes it possible to reuse and share information. We shall look at these changes in the following sections.

10.4 NEW SOURCES OF DATA

It is many years since one of the authors (David Green) first went walking through the forests of eastern Victoria, in Australia's south east. In all that time, the forests hardly seem to have changed at all, and yet in many places, you will see huge white skeletons of old trees sticking out above the canopy. These skeletons are the remains of trees that were killed by the huge bushfires that devastated the entire region in the year 1939. The skeletons stand out because in all the years since then, the forests have still not grown back to their former condition.

This story highlights the slow pace of ecological change, which often occurs on a vast scale, both in time and space. To go out and observe nature at one time and place can give a false impression of what is really going on. To come to grips with the complexities wrought by ecological interactions, we therefore need to be able to observe ecosystems, not only here and now, but also over large regions and over large periods of time.

One of the benefits that technology has brought to landscape ecology is the ability to expand the scale on which we can observe ecosystems. Some of the most basic questions in landscape ecology include where are organisms found? In what numbers? And why? In the past few decades, many new technologies have appeared that help us to answer such questions. Automated weather stations, satellite imaging systems, animal tracking and physiological monitoring equipment have all enabled us to collect more and better data on ecological systems. These technologies have also brought to landscape ecology the ability to monitor ecosystems on a more or less permanent basis, rather than intermittently over short periods.

More recently, the combination of geographic information systems (GIS), satellite remote sensing systems and global positioning devices have revolutionised the problem of dealing easily with the huge amounts of spatial data that can now be sensed and captured. Visible, infrared, near infrared, thermal, radar, ultraviolet, gamma, electron and neutron radiation can all be used to measure different processes and states in landscape systems.

Airborne sensing of natural gamma emission by phosphorus, uranium and thorium has led to a revolution in soil mapping, providing not only improved mapping of observed vegetation patterns, but also identification of underlying geological patterns, all in a fraction of the time and cost associated with a conventional soil survey. Unexpected benefits of this new technology have been the ability to rapidly map areas of extreme soil pH and salinity, and to identify causes of low plant productivity. This mapping is so successful that it is becoming difficult to sell land for cropping or grazing without submitting it to such an analysis.

Sometimes the application of new technology to existing problems consumes huge amounts of resources but yields nothing of value. For example, much work has been done on carbon isotope ratios in plant tissue, trying to relate these to plant water relations. Year after year, more work is reported and more resources have been consumed, but alas, no consistent results have been found. Sometimes, putting fascination with new technology ahead of the solution of ecological problems can result in enormous waste.

10.4.1 Putting IT to work

A good example of the application of GIS and ecological models to conservation and environmental management was the Everglades Landscape Model (ELM) developed by a consortium of agencies lead by Constanza (Fitz et al. 1996). ELM was a regional scale ecological model designed to test different water management scenarios in southern Florida, where there was conflict between conservation of the Everglades and the needs of agriculture, mainly the sugar industry, as well as residential developments along the Atlantic coast, including Miami.

ELM comprised models of the hydrology, soil and water nutrient flows, aquatic flora and fauna and the vegetation biomass and community types of the Everglades region. The region was divided into 10,000 one square kilometre grid cells. Hydrological, nutrient and biological information were stored for each cell. In the model, vegetation responded to the conditions in each cell. Movement of water and nutrients between cells was simulated according to different management scenarios. The output comprised varying patterns of plant community distributions.

For example, the establishment of a network of canals and levees in the early 20th century to reduce flooding and to provide water for irrigated agriculture and residential use dramatically altered flows through the Everglades (Costanza 1975). Reduced fresh water flow into Florida Bay resulted in salt water incursion in the southern Everglades with consequent changes to the distribution of mangrove species and threats to species such as the manatee, which all but disappeared.

10.4.2 Serendipity

As we saw at the beginning of this chapter, it is commonplace in scientific research to find that the application of new technologies leads to new discoveries. The ability to measure very low concentrations of abscisic acid in the transpiration stream of plants led to the discovery in the early 1980s of signals passing directly from the root tips to the stomata. When root tips experience moisture levels below a critical level – enough to satisfy evapo-transpirative demand upon the plant foliage – abscisic acid is produced in low concentrations, and when this reaches the stomata, they close. Of course, under more extreme conditions, higher concentrations result in leaf abscission. This explains two primary responses of plants to increasing soil water deficits. Prior to this discovery, the prevailing explanation for the passage of water through the soil-plant-atmosphere continuum was that of hydraulic theory, which never managed to explain all the results of

Take any issue in landscape ecology or environmental management. How will cutting a new road through a forest affect the biodiversity of the area? Will land-clearing put any rare species of plants or animals at risk? How much land needs to be set aside to ensure connectivity of a species' range? Each of these questions requires information. To answer any one of them, not to mention hundreds of others, you need information that is accurate, correct, reliable, up-to-date and comprehensive. But that is not all; you also need the ability to combine different kinds of information, possibly from different sources, and to interpret that information in appropriate ways.

One of the challenges for environmental information management has been to create systems that can deliver rapidly and cost effectively exactly the information that people need. To deal with complex issues, that information needs to be integrated with appropriate tools for data analysis, simulation modelling and visualisation.

10.6.1 The key questions

Information management is a compromise between what should be done and what can be done with available resources. Many constraints limit the effort that can be devoted to the development and maintenance of conservation information, so these systems are usually designed to address issues of the highest priority, generally to answer conservation questions that are important and urgent.

Conservation managers need answers to many questions. Primary questions are concerned with patterns, such as where a particular species is to be found. Secondary questions are concerned with process, such as what factors influence where a species is found. Secondary questions often involve relationships between different patterns, such as whether the distribution of a species overlaps conservation areas. Most conservation questions are inherently geographic in nature. They deal with a range of issues, but in general, they concern several essential elements, or 'layers'. Primary questions normally concern issues within a single layer, whereas secondary questions concern interactions between layers.

10.6.2 Modelling species distributions

One application for GIS is in modelling the distribution of species. Relating known locations of species to values of environmental variables for those locations leads to models of the way species distributions are influenced by various environmental factors. The classic approach to this problem has been classification, which places locations or regions into categories. The distribution of (say) rain-forest tree species would be expected to coincide with areas mapped as rain forest. However, the use of such categories can be unreliable, as species tend to be plastic in their associations.

An alternative approach is to associate species locations with environmental variables directly (Boston and Stockwell 1995). For instance, the programs BIOCLIM (Busby 1991) and CLIMEX (Sutherst et al. 1999) relate distributions of plants, animals and insects to a climatic range. In each case, applying the models to the entire landscape identifies all sites that lie within the envelope. This approach can also be used with relevant climate change scenarios to predict the potential effects on plant and animal distributions. This form of modelling has proved to work well in some regions, such as Australia, but assumes that distributions are in equilibrium with the environment. The models do not necessarily predict the true distribution, only the potential distribution. For instance, competition can restrict a species' distribution to a narrower range than their potential. For this reason, the method is seen as more problematic in the chronically disturbed environments of Europe.

10.6.3 The role of the Internet

The rise of the Internet, and especially the World Wide Web, revolutionised the dissemination of biodiversity information during the 1990s. Once an organisation had published information online, anyone, anywhere could access it. This capacity meant that people could access relevant information more simply than in the past. They could also access it faster and in greater quantities. As well, it raised the potential for organisations to provide wider access to information resources that normally require specialised software or hardware. For example by filling in an online form, remote users can query databases. In many cases, geographic information that previously required specialised (and often expensive) equipment can now be accessed and viewed remotely via a standard Web browser Green and Bossomaier 2002).

The explosion of online information is a problem: finding one item amongst millions is akin to finding a needle in a haystack. Potential solutions for searching include the use of intelligent agents that continually sift and record relevant items, and the promotion of metadata standards to make documents self-indexing. Information networks (see below) provide a way of organising sources of information.

10.6.4 Practical issues

Many attempts at large scale information sharing of environmental information have failed because of issues arising over ownership, control and the cost of data. However, the Internet makes it possible for agencies to share data while retaining control over it. As a commodity, environmental data used to be rare and expensive. Now it is becoming abundant and cheap. An on-going issue in environmental information has been what commercial model to use when sharing information. Charging for datasets has been a crucial source of revenue for many organisations, but this practice inhibits its widespread use. One school of thought is that (like genomic data) the benefits of freely sharing environmental data outweigh any short-term losses. However, a vexed issue is how to compensate the contributors of data to commercial products. An example would be an online resource map generated from databases that are maintained at other sites.

10.6.5 Many hands make light work

In ecology, people collect data in research projects for many different reasons over a long period of time. So the process of assembling particular data for a new purpose from multiple sources is required before queries and analysis can begin. However, there are some common ways of organising data, such as geo-referencing, which organises data in relation to latitude, longitude, elevation and time of collection. Geo-referencing makes for ready access using geographic information systems. The search engines that enable us to find answers to queries from the petabytes of information on the World Wide Web can deliver those answers through the processes of systematically harvesting information and organising it into indexes. These indexes are not unlike those used in any library, except that they are dealing with a larger and more diffuse collection of information sources.

Many carpenters have learned to their cost that half inch nuts cannot be used with one centimetre bolts. Tools manufactured in imperial and metric sizes are incompatible with one another. In the same way, datasets from different sources cannot be combined unless they are compatible. To be reusable, data must conform to standards. The need for widely recognised data standards and data formats is therefore growing rapidly. Given the increasing importance of communications, new standards need to be compatible with Internet protocols.

Four main kinds of standards and conventions are commonly used:

1. Information design standards and information models describe in conceptual terms the information needs of an enterprise. All data and information are collected, stored and disseminated in the framework.
2. Attribute standards define what information to collect. Some information (for example, who, when, where and how) is essential for every data set; other information (for example, soil pH) may be desirable but not essential.
3. Quality control standards provide indicators of validity, accuracy, reliability or methodology for data fields and entries.
4. Interchange formats specify how information should be laid out for distribution (Figure 1-4).

For example, the Species 2000 project defined the following seven standard items for biodiversity data: accepted scientific name, synonyms, common names, latest taxonomic scrutiny, source database, comments, and family.

In general, datasets need to be accompanied by metadata – data about the data. Metadata provide essential background information about datasets, such as what it is, when and where it was compiled, who produced it, and how it is structured. Without its accompanying metadata, a dataset is often useless. Metadata have gained considerable prominence as indexing tools, especially since the advent of large-scale repositories on the Internet. Because of the vast range of information online, the W3C Consortium espoused the principle that items should be self-documenting. That is, they should contain their own metadata.

Whatever the material concerned, its metadata always need to cover the basic context from which the information stems. Broadly speaking. metadata need to address the following basic questions:

- HOW was the information obtained and compiled?
- WHY was the information compiled?
- WHEN was the information compiled?
- WHERE does the information refer to?
- WHO collected or compiled it?

The Dublin Core (DC), designed originally for online library functions, specifies a suite of fields that should be used in identifying and indexing Web documents (Weibel et al. 1998). The fields are: title, creator, subject, description, publisher, contributor, date, type, format, identifier, source, language, relation, coverage, rights. Two other important standards are XML and RDF. The Extensible Markup Language (XML) makes metadata an integral part of the organisation and formation of documents and data (W3C 2005). The Resource Description Framework (RDF) provides a general approach to describing the nature of any item of information (Lassila and Swick, 1998).

All the above makes technical sense, but people have to agree to adopt compatible standards. This is a role that national and international organisations are increasingly taking over, as we will see below (Section 10.7).

10.6.6 Quality not quantity

Early in 1991, the Australian Government completed a project to determine the distributions of rare and endangered species. The data were collated from many different sources, from studies carried out by many different people over many decades. When all the data were finally available, they were plotted on a map of the entire continent. What emerged was a complete map of roads in central Australia. What this indicated was a systematic bias in the way in which people had carried out their field studies. Although each individual study may have been planned and well executed, and although they may have delved into previously unknown territory, they were all constrained by the practicalities of accessing remote areas. The result was that areas relatively close to roads were the most frequently studied; in some regions they were the only areas studied. So taken over all, there were large gaps in the coverage and bias towards areas close to human activity.

Even more disturbing was the fact that on the map many sites were found to lie out in the ocean. How could trees that normally grow in mountain forests or on plains exist far out to sea. Obviously they could not. It was clear at once that the site records plotted in the ocean all contained errors. Even more disturbing was the implication that many other records, ones not obviously wrong, must also contain errors.

The above example highlights a crucial issue in compiling and managing data: quality. Quality is a prime concern when compiling information. Incorrect data can lead to misleading conclusions. They can also have legal implications. The aim of quality assurance is to ensure that data are valid, complete, and accurate, and this is necessary for data records to be useable. They must also conform to appropriate standards so that they can be merged with other data. Errors in any field of a data record are potentially serious.

The most direct method of assuring quality is to trap errors at source. That is, the workers recording the original data need to be rigorous about the validity and accuracy of their results. If electronic forms are used (for example, over the World Wide Web), then two useful measures are available. The first is to eliminate the possibility of typos and variants by providing selection lists, wherever possible. For instance, selecting (say) the name of a genus from a list eliminates miss-spelling (though selection errors are still possible). For free text, scripts can be used to test for missing entries and for obvious errors prior to submission. Online forms, for instance, can automatically test whether a name entered is a valid taxonomic family or whether a location falls within the correct range (Green and Bossomaier 2002).

10.6.7 The ERIN project

This task of providing sufficient leadership to convince many different institutions to share data was faced in the development of the Environmental Inventory Resource Network (ERIN) during the early 1990s. The motivation for that project was that many policy decisions faced by all levels of government could be better answered when spatial data was taken into account. Historical data had to be either gathered or linked for access and queries, and then some of the new computational methods that are the topic of this book had to be employed to make predictions. Being able to visualise in a GIS, for instance, the consequences in ten years of a change to policy has proved of enormous benefit. For example, it is possible to visualise maps of vegetation showing the consequences of restrictions to old growth forest harvesting. On the landscape scale, studies of biodiversity and biogeography require detailed data about both environmental conditions and species distributions.

Continental scale summaries available through data warehousing agencies such as ERIN show that Australia has made great progress in compiling environmental information (Green 1992). However, the task is so vast that only by mobilizing the resources of every relevant agency and research group is it possible to compile even a moderately adequate coverage at scales suitable for policy decisions. To put the size of the task into perspective, consider retrospectively digitizing the national insect collection of over eleven million specimens. At an annual rate of 50,000 specimens, it would take over 200 years (Richardson and McKenzie 1992).

10.7 BIODIVERSITY INFORMATION

It is easy to be misled into thinking that local patterns are global. Introduced plants may be common along roadsides, but that does not necessarily mean that they pervade the entire forest. On the other hand, seeing remnants of native vegetation by the roadside might give motorists a false sense that native species and ecosystems are not under threat. Likewise, as we have seen, a little information can fool people into thinking that they have the whole picture.

The rapidly growing scale of environmental alteration and increasing public awareness of environmental issues has highlighted a need for off-reserve conservation and for broad-scale landscape management. Such issues include: environmental impact assessment, state of the environment reporting, environmental monitoring, conservation of rare and endangered species, natural heritage planning, species relocation programs, land use planning, and environmental restoration. Out of such activities, it has become increasing clear that local decisions and priorities need to be set in a wider and ultimately global context.

Stimulated by the rapid growth of the World Wide Web, the 1990s saw a proliferation of international projects that aimed to co-ordinate the compilation of biodiversity information. A useful outcome of networking activity has been to put much biodiversity information, such as taxonomic nomenclature and species checklists, online. One of the first priorities was to develop consistent reference lists of the world's species. For instance, the International Organisation for Plant Information (IOPI) began developing a checklist of the world's plant species (Burdet 1992). The Species 2000 project has similar objectives (IUBS 1998). At the same time, the Biodiversity Information Network (BIN21) set up a network of sites that compiled papers and data on the biodiversity of different continents. There are now many online information networks that focus on environment and resources (Table 1).

The greatest challenges in collating biodiversity information have been human, especially legal and political issues, rather than technical problems. One outcome of the 1992 Convention on Biological Diversity was agreement on the concept of a Clearinghouse Mechanism, under the United Nations Environmental Programme (UNEP). This scheme aimed to help countries develop their biodiversity information capacity. The longer-term goal was to enhance access to information by promoting the development of an international system of clearing houses. These sites gather, organise and distribute biodiversity information.

In 1994, the OECD set up a Megascience Forum to promote large science projects of major international significance (Hardy 1998). The Human Genome Project was one such enterprise. Another was the proposal for a Global Biodiversity Information Facility (GBIF). The aim of GBIF is to establish

"... a common access system, Internet-based, for accessing the world's known species through some 180 global species databases ..."

Primary industries set up similar initiatives. For instance, in 1996 the International Union of Forestry Research Organisations established an international information network, and in 1998 began work to develop a global forestry information system (IUFRO 1998).

All of these developments point to the increasingly global nature of conservation and environmental management. In the final Chapter, we look at landscape ecology from a global perspective, and especially some of the complexities that human societies introduce.

*Table 10-1 Some environmental organisations and services online *.*

Organisation	Web address
Australia-New Zealand Land Information Council (ANZLIC)	http://www.anzlic.org.au/
Australasian Spatial Data Directory	http://www.environment.gov.au/net/asdd/
Center for International Earth Science Information Network	http://www.ciesin.org/
Convention on Biological Diversity	http://www.biodiv.org/
CSU Mapmaker	http://life.csu.edu.au/geo/
Diversitas	http://www.icsu.org/DIVERSITAS/
Environment Australia	http://www.ea.gov.au/
Environment Australia interactive maps online	http://www.ea.gov.au/epbc/interactivemap/
European Environment Information and Observation Network (EIONET)	http://www.eionet.eu.int/
Global Biodiversity Information Facility (GBIF)	http://www.gbif.org/
International Legume Database and Information Service (ILDIS)	http://biodiversity.soton.ac.uk/LegumeWeb
International Organization for Plant Information (IOPI)	http://life.csu.edu.au/iopi/
International Union of Forestry Research Organizations (IUFRO)	http://iufro.boku.ac.at/
Open GIS Consortium	http://www.opengis.org/
Species 2000	http://www.species2000.org/ http://www.sp2000.org/
The TIGER mapping system.	http://tiger.census.gov/
United Nations Environment Programme (UNEP)	http://www.unep.org/
USDA PLANTS Database	http://plants.usda.gov/
US National Geophysical Data Center	http://www.ngdc.noaa.gov/paleo/
US Satellite Active Archive	http://www.saa.noaa.gov/
World Conservation Monitoring Centre (WCMC	http://www.unep-wcmc.org/
World Wide Web Consortium.	http://www.w3c.org/

* This list is far from exhaustive. Web addresses change frequently. The addresses given here were correct at the time of publication.

CHAPTER 11

THE GLOBAL PICTURE

Eroded desert landscape near Mount Sinai, the legacy of many millennia of human occupation.

The acid test for any idea is whether it works in practice. As we have seen throughout this book, complexity theory teaches us many lessons that have immediate practical uses in landscape ecology, in conservation and in environmental management. However, if complexity theory teaches us any lessons at all, it teaches that we cannot understand ecosystems by studying them in isolation. To get the full picture, we need to set each ecosystem in the context of its surrounding region. Nor can we understand an entire region without understanding how the elements in its ecological mosaic interact with one another.

No ecosystem is a closed box. Whether it is a forest, a mountaintop, or a swamp, every ecosystem exists within the larger context of the surrounding region. At first glance, a lake or pond may seem to be isolated from its surroundings, but it is not. It interacts with the outside world. Water flows in from surrounding sources. It carries with it nutrients, living organisms, debris and other material, all of which can affect the pond's ecosystem. Birds come and go. Maybe they catch fish, or frogs or insects in the pond. Sometimes they bring material with them from other ponds. This may include eggs, seed or larvae of foreign species.

The region within which an ecosystem is found influences local events in many ways. The physical environments interact. Air and water move back and forth. Plants and animals move between the ecosystem and the surrounding region. The same holds true on ever larger scales. Ultimately, every ecosystem needs to be considered not only within its regional context, but also its continental and even global setting.

As well as the geographic context, ecosystems and landscapes are bound by the context of their history. Recent years have seen many discoveries about the long-term role of human disturbance in shaping environmental history. These findings provide some worrying lessons, as we will see in the following sections. There we will look more closely at some of the ways in which human activities impinge on environments and on landscape ecology.

In today's world, if there is any context in which we should be setting practical matters involving landscape ecology, then it is surely the relationship between environments and human activity. As we saw at the very beginning of this book, major conservation issues, such as the clearing of the Amazon rain forest, are often intricately bound up with social, economic and political issues. In this final chapter, we will look briefly at some of the large scale complexities introduced into landscape ecology by the growth of human societies.

11.1   HUMANS AND THE GLOBAL ENVIRONMENT

In Chapter 8, we saw how environmental disasters have shaped the evolutionary history of life on Earth. Considering this, it is not surprising that people have wondered whether cataclysms have played any part in human history. During the 1990s a spate of books appeared on this topic. For instance, dendrochronologist Mike Baillie argues in his book *Exodus to Arthur* that comets caused several disasters that affected historical events over the last five thousand years (Baillie 1999). In his 1999 book *Catastrophe*, British author David Keys argues that an environmental disaster in the year 535 AD, possibly a major eruption by Krakatoa, had global repercussions (Keys 1999). It darkened the sky for some 18 months and triggered a long series of social and political events. The consequences, he claims, include the first recorded outbreak of bubonic plague, widespread famine, and the fall of the Roman Empire.

Palaeo-environmental studies, especially pollen records, are beginning to reveal just how great an effect environmental change has had on human history. Previously, human affairs had been viewed as independent of nature. Increasing attention to environmental factors perhaps stems from growing concern about current environmental problems. In this respect, history provides a salutary lesson about the dangers of environmental degradation.

Early human societies consisted of nomadic tribes who hunted and gathered food. Around 10,000 years ago, agriculture arose, permitting the formation of large permanent settlements with centralised governments—in other words, civilisations appeared. These civilisations rapidly grew in size and power, displacing less technologically advanced native peoples worldwide. Today, the few hunter-gatherer societies that remain are descendants of these native peoples.

11.1.1 The fall of civilisations

Palynological and archaeological evidence show that the fall of many past civilisations was at least partly a result of over-exploitation or environmental change. The sad lesson of history is that most ancient civilisations proved environmentally unsustainable. Many civilisations over-exploited the landscape, so wiping out the natural resources on which their existence depended. These civilisations ultimately suffered the consequences of landscape degradation. It may be no coincidence that the lands of the Middle East, regarded as the cradle of civilisation and once famed for their agricultural wealth, are now deserts. If modern western civilisation appears to have escaped this fate, then it should be remembered that over the past few hundred years, western nations have coped with the problem only by extensive use of fertilisers and artificial nutrients, and by extending their reach via empires that cover the globe.

In his 1991 book *A Green History of the World*, author Clive Ponting describes the ways in which several ancient civilisations, and other communities, so altered their environment that it could no longer support them (Ponting 1991). A case in microcosm was Easter Island, where the islanders apparently removed all the trees in the course of building their famous statues. The loss of the trees led to erosion and most bird species became extinct. Without trees, no canoes could be made for fishing. The consequent famine killed three quarters of the human inhabitants and the island has never recovered (Diamond 2000). For civilisations such as the Easter Islanders, the Mayas and the Sumerians, cultural collapse followed as a direct result of environmental degradation.

The case of the ancient Sumerians provides a salutary lesson for modern times. Here was a magnificent empire that at its height encompassed many thousands of square miles of territory. Ponting (1991) argues that the development of their empire forced the ancient Sumerians to consume resources at an unsustainable rate. It took about 1000 years for this process to run its course. Although the

transformed the ecology of the region: around the time when humans colonised Australia and the megafauna vanished, there was a drastic and permanent reduction in the diversity of plant foods available to birds and animals, along with increases in fire-adapted plant communities (Miller et al. 2005).

Curiously, Africa does not appear to have suffered from the Pleistocene extinctions. The difference in Africa seems to have been that humans evolved there (Flannery 2001). So animal species living in the same ecosystems had time to adapt to human activities, and humans had time to adapt and to co-evolve sustainable interactions with the plants, the wildlife, and the landscape.

11.1.3 The current ecological crisis

Although we may never know exactly what role humanity played in the Pleistocene extinctions, the evidence suggests a harsh warning for modern civilisation. It seems that it has always taken humans a long time to develop sustainable practices for exploiting natural and living resources, and the process routinely devastated the pre-existing ecology. Moreover, sustainable societies have arisen not through human ingenuity, but through ecological feedback processes that have removed excess humans, combined with the loss of ecologically incompatible species. Our mere existence has already changed the world's ecosystems irrevocably. Restoration of a 'natural' global ecology is therefore impossible. What we must aim for instead is a sustainable way of life.

In the modern world, we do not have the millions of years that it took for humans to evolve in Africa. We do not even have the thousands of years that it took native peoples to learn how to manage new lands after they migrated into them. The pace of ecological change, and the worldwide scale on which it now occurs, are unprecedented. At the very most we have perhaps a couple of generations in which to adapt our technologies, our economies and our culture.

Stories of modern environmental catastrophes brought on by human activity are legion. Although extinctions are an inevitable part of life on Earth, never in the history of the Earth has one species been so clearly responsible for rapid global change as in the last 100,000 years with *Homo sapiens*. Current extinction rates are believed to be thousands of times greater than the normal rate throughout evolutionary history (Wilson 1988). Based on habitat loss alone, it has been estimated that about 30 per cent of living species will be extinct by the middle of the twenty-first century (Novacek and Cleland 2001), while half of all species are expected to become extinct during the next century (McKinney and Lockwood 1999). These estimates do not take into account the effect of climate change, which itself is expected to drive 15-37 per cent of species to extinction by 2050 (Thomas et al. 2004). Given these statistics, the current extinction event is similar in scale to the five greatest mass extinctions in the history of complex life. For this reason, Richard Leakey called the current era the "Sixth Extinction" (Leakey 1994).

In 1984 Peter Vitousek and colleagues attempted to quantify the global impact of human activity (Vitousek et al. 1986). They estimated that about 40 per cent of the total energy generated by photosynthesis in terrestrial ecosystems was co-opted by humans. Humans have doubled the input of nitrogen to terrestrial ecosystems, causing depletion of essential soil nutrients and acidification of soils and aquatic systems, thereby damaging many aquatic ecosystems (Vitousek et al. 1997a). Phosphorus storage in both terrestrial and aquatic ecosystems has increased by at least 75 per cent during industrial times, contributing to lethal algal blooms (Bennett et al. 2001). In 2002, ecologists assessed the sustainability of total human usage of the biosphere (Wackernagel et al. 2002). They found that even in 1961, humans had monopolised around 70 per cent of the biosphere's total productivity. By 1999, this figure had grown to 120 per cent, overshooting the maximum sustainable usage by 20 per cent.[1]

The problem is that we are just beginning to appreciate the complex nature of plant and animal populations and the ways in which they interact with each other and with the environment. As often as not, past efforts to control and manage ecosystems have been disastrous.

Lever (2001) relates the graphic story of the cane toad (*Bufo marinas*) in Australia. In the early 1930s, Australia's sugar crop was under threat from the cane beetle (*Lepidiota frenchi*). After studying the options, the Queensland government decided to introduce a predator, the cane toad. The cane toad was known to eat beetles and the crops of Hawaii were free from the beetle problem. However, ecologists failed to take into account the cane toad's adaptability. Once they were introduced, the animals ate the cane beetles. But they also ate anything else that would fit into their mouths. They immediately began to massacre the small animal population of the region and the problems did not stop there. From two glands on each side of the back of its head, the cane toad exudes poison as a defence against predation: any animal that tried to eat it died. With ample supplies of food, and no natural enemies, the cane toads began to spread. Today they are a huge environmental menace in Australia, ranging from northern New South Wales to the Northern Territory.

Cane toads have prospered through human intervention, but the same cannot be said for many other amphibians. All over the world, frogs and salamanders of a wide variety of species are becoming rarer (Alford and Richards 1999). For many years, conservationists searched for a common disaster that would explain this widespread trend. But none has been found. Instead, it is increasingly evident that the demise of the amphibians is caused by many different stressors simultaneously (Blaustein and Kiesecker 2002). Such stressors include habitat destruction, increased ultraviolet radiation due to destruction of the ozone layer, climate change, pollution, disease, and the introduction of alien species. Virtually all of these events can ultimately be traced to human causes.

Many of these stressors affect other kinds of organism. Why should amphibians be more affected globally than other groups? The underlying reason seems to be simply that amphibian populations are unusually vulnerable to any sort of stress. Most amphibians require both aquatic and terrestrial homes. They

---

[1] This is possible because the biosphere also includes stored value, or "environmental capital", such as forests, fisheries and coal reserves, that are the outcome of long-term accumulation. We are rapidly eating into these reserves.

have poor dispersal abilities and are exceptionally vulnerable to pollutants and infections because of their moist, absorbent skin. Blaustein and Kiesecker (2002) point out that if there is a single common factor in amphibian decline, it is that the causes are complex and involve both global and local processes: "One consistent theme appears to be the interactions between environmental change at local (e.g. habitat modifications), regional (e.g. acidification or contaminants) and global scales (e.g. climate change or UV-B radiation) with the modification of local biotic interactions (e.g. disease or introduced species)".

11.2   GLOBAL CLIMATE CHANGE

11.2.1 The runaway greenhouse

One of the greatest fears regarding atmospheric pollution is that it will lead to a runaway global greenhouse effect. A dramatic, practical demonstration of the impact that an extreme greenhouse effect can have is provided by the planet Venus (Kasting 1988). Theories of the origin of the solar system suggest that Venus, Earth and Mars are all composed of similar materials. It is now thought that the atmospheric composition of Venus and Mars, both of which are largely carbon dioxide (96 per cent) and nitrogen (~3.5 per cent), reflect the early composition of Earth's atmosphere. The present high concentrations of oxygen and nitrogen are thought to be a by-product of photosynthesis and other organic activity. Although Venus is almost a twin of the Earth in size, their surface conditions could hardly be more different.

For centuries, Venus - the Morning and Evening Star - was thought of as a heavenly place. In reality, it resembles hell. The surface temperature (464°C) is hotter than an oven; the air pressure is the same as water pressure at a depth of 1 kilometre under oceans on Earth; and the atmosphere is laced with sulphuric acid and other toxic gases. The few spacecraft that have landed there have all been cooked, crushed or corroded within minutes of their arrival. These dreadful conditions are now seen to be the result of a runaway greenhouse effect (Ingersoll 1969). The clouds that for so long shrouded the surface of Venus in mystery act as a blanket that keeps in the heat from the sun. As in a greenhouse, the heat becomes trapped, keeping the surface intolerably hot.

On Earth, a relatively weak natural greenhouse effect maintains temperatures hospitable to life (Kasting et al. 2001). Human activity is greatly increasing levels of greenhouse gases such as carbon dioxide and methane. Some scientists fear that if greenhouse gases accumulate beyond a critical point, they may become part of a positive feedback loop, with greater temperatures increasing the concentration of greenhouse gases, which in turn lead to still greater temperatures, and so forth. Such positive feedbacks could involve several processes.

Evaporation resulting from warming increases water vapour, which acts as a greenhouse gas (Held and Soden 2000). Ice is more reflective than earth or sea, so melting icecaps, glaciers and snowfields could accelerate global warming. Marine plants and animals store carbon in their tissues, and some of that carbon falls to the ocean floor when these organisms die. Consequently, the ocean biota act as a highly effective carbon sink (Siegenthaler and Sarmiento 1993), but their carbon absorption is expected to decline as the oceans grow warmer (Chuck et al. 2005). Within limits, increasing carbon dioxide leads to increased photosynthesis, so that carbon is stored by plants. However, as lands become warmer, plant and soil respiration increase, releasing more carbon dioxide (Cox et al. 2000). Rainfall may decline in many areas as a result of warming, leading to accelerating loss of carbon-storing vegetation (desertification), and consequent increases in atmospheric carbon dioxide (Millenium Ecosystem Report 2005). Warming of the ocean floor could also lead to the release of methane stores, further increasing global warming (Brewer 2000). Researchers have suggested that a runaway greenhouse effect of this type at the end of the Permian era caused a mass extinction in which 95 per cent of species died out (Benton and Twitchett 2003).

Unfortunately, because of the numerous non-linear factors involved, predicting exactly when and how any of the above factors will influence climate change is impossible. As we saw in Chapter 5, even systems with just three factors and near-perfect data can rapidly become unpredictable.

Indeed, for many years one school of thought in this controversial area believed that atmospheric pollution would lead to global cooling. Proponents argued that clouds reflect heat, thus reducing surface temperature. This effect underlies concerns about the so-called 'nuclear winter' that could follow atomic warfare. It does seem to have occurred many times in the past. For instance, if David Keys is correct, then the eruption of Krakatoa in 535AD, described above, had exactly this effect. The climatic disruption lasted for years, disturbing ecological systems and causing enormous long-term repercussions for world politics.

11.2.2 Is global warming happening?

The controversial debate about global warming highlights the problems that can arise when confronted by complex environmental phenomena. The confusion in models has been matched by the lack of clarity in the data. The essential problem in trying to demonstrate global climate change is to show that recent changes go outside the norms. But what exactly are those norms? In the past, climates have changed dramatically over time, and so most modern variations fall well within the limits of past cycles. There is also the question of scale. Systematic, instrumented weather records date back only 150 years at most, so it is difficult to show statistically that recent variations fall significantly outside the modern range (Overpeck 2000). Palaeoclimatic records, such as ice cores and tree rings, are needed to demonstrate the significance of recent trends.

For this reason, the pattern of global climate change has been difficult to prove. For many years, sceptics remained unconvinced of any change. Today, the evidence for global warming is accepted by the overwhelming majority of scientists and by most governments (Kerr 2001; Vinnikov and Grody 2003). In a recent study, for instance, Huang and his colleagues obtained records of ground surface temperature dating back about 500 years from hundreds of cores on every continent except Antarctica (Huang et al. 2000). The results of this and other studies imply that the 20th century is the warmest period in the last 500 years, by about 0.5°C (Intergovernmental Panel on Climate Change 2001). The warming also occurred faster than any other century during that period (Overpeck 2000).

On the 16th of February, 2005, the United Nations' Kyoto Protocol, intended to limit greenhouse gas emissions, finally came into force after seven years of dispute. Studies suggest that government measures to cut emissions remain inadequate to curb climate change; but many governments and industries oppose further reductions because of their likely economic impact (Nordhaus 2001). Although global warming is widely accepted, some sceptics still question whether human emissions are responsible for global warming, and whether the warming is a problem.

Resolving the issue is one of the most urgent questions of our time. Without absolute proof that damaging climate change is resulting from greenhouse gases, nations and industries see no reason to change. Why cut carbon emissions if they do no harm? However, if fears of runaway greenhouse effects are correct, then by the time the evidence becomes irrefutable, it may already be too late to halt the process. Already, human health impacts may be evident: for example, one study claims that global warming almost certainly doubled the likelihood of extreme heatwaves such as that which killed 20,000 Europeans in 2003 (Stott et al. 2004).

Scenarios such as this have led environmentalists to advocate the *precautionary principle*, which essentially states that when faced with uncertain risks, actions should be forbidden until proven harmless (Kriebel et al. 2001). The precautionary principle is highly controversial. Critics argue that application of the principle will stifle scientific and technological advances, because it is virtually impossible to demonstrate that anything is completely safe (Sunstein 2003). Balancing these concerns in applying the precautionary principle remains one of the greatest challenges for the global environmental movement.

11.2.3 Ecological effects of global warming

Given the strong evidence for global warming, the crucial question for conservation is whether changes are occurring in the distributions of ecosystems and individual species. Studies suggest that the distributions of a wide range of animals have already changed in response to global warming (Kerr 2001). Evidence for the ecological impact of climate change was amassed by collaboration between ecologists from Europe, Australia and the United States (Walther et al. 2002). They concluded that "there is now ample evidence that recent climatic changes have affected a broad range of organisms with diverse geographical distributions".

According to Walther et al. (2002), many species are beginning spring activities early. Birds are migrating, singing and breeding earlier. Butterflies are maturing earlier. Frogs are singing and spawning earlier. Plants are shooting and flowering earlier. Simultaneously, autumn behaviours are occurring later in the season: migrations are delayed, and leaves are changing colour later. Many of these seasonal activities are directly cued by physiological effects of temperature, so the shifts in time are not surprising. There are, however, strong variations between species in the way in which they respond to these changes.

In addition, Walther et al. reported that a wide variety of species have shifted their geographic ranges poleward in response to recent climate change. This has resulted in invasions by exotic species, such as warm-water species in the Mediterranean Sea. In Antarctica, mosses, higher plants, and soil invertebrates are spreading rapidly. Coral bleaching events, which involve massive local mortality of coral and are attributable to high temperatures, are increasing in both frequency and severity. At the community level, changes in the relative abundance of different species, such as increases in woody shrubs in the United States, are attributed to climate change.

It remains unclear, however, how all of these observed changes will combine to affect complex ecosystems. One of the most worrying patterns is the desertification of the dryland ecosystems. These cover 41 per cent of the earth's surface and include 44 per cent of its croplands (Millennium Ecosystem Assessment 2005). As we saw in Chapter 5, stability analysis suggests that dryland ecosystems are unstable, liable to collapse to desert when faced by relatively small perturbations due to positive feedback effects. The Millennium Ecosystem Assessment reported in 2005 that drylands were home to 2.1 billion humans – one third of the earth's population – among them many of the world's poorest peoples. Around 10-20 per cent of drylands showed deteriorating productivity due to encroaching desertification. Climate change is expected to decrease rainfall in most drylands, exacerbating desertification and further increasing net carbon emissions due to vegetation losses. This may already be occurring: some authors argue that global warming has contributed to the worst Australian drought in history (Karoly et al. 2003). However, in other areas increasing carbon dioxide may, within limits, enhance plant growth, to some extent offsetting this effect. The response of individual ecosystems may therefore be highly variable.

11.3   GLOBALISATION

One of the most controversial trends at the end of the 20th century was for organisations and processes to grow until they become worldwide. This tendency, usually called *globalisation*, means that the interactions between environment, resources, economics, and politics are rapidly increasing. Communication is one of the prime factors that make globalisation possible.

In globalisation we see a classic example of the big getting bigger and the rich getting richer. Driven by the commercial imperative of continual growth, companies are always seeking to expand and grow. A positive feedback loop operates, in which the bigger and more successful an organisation becomes, the greater is its ability to grow still further. So really successful companies grow and grow, acquire more and more assets, expand into new countries, and eventually become multinational conglomerates.

By the turn of the millennium, globalisation had become an issue that was generating heated debate all around the world. Many would argue rightly that globalisation was not new. Governments, companies, and other organisations have spread their influence internationally for centuries. Nevertheless, modern transport systems and information technology have increased the rate of movement of money, goods, people, and ideas astronomically. Historically, economic, social, and environmental changes within one community would have impacts on the immediately surrounding ones, ultimately rippling to the global level, but these impacts were likely to be reduced at every step by the relatively insular dynamics of the individual communities. Today, communities everywhere are so tightly interdependent that a change in one is likely to be felt immediately by all. This observation applies as much to environmental issues as to any others.

One of the most outspoken critics of globalisation is Jerry Mander, a senior fellow of the non-profit advertising agency the Public Media Center. In a recent interview, Mander (London 2000) argued that

> *"Globalisation doesn't work... We're on the verge of an ecological catastrophe of stupendous proportions."*

In Mander's view, globalisation is distinct from capitalism in that globalisation leads to power being concentrated in giant corporations:

> *"the great part of the system doesn't function in a capitalist manner. It's not a socialist manner either. It's some kind of hodge-podge of connections that have been put together for greasing the skids of advanced development and growth and corporate benefit... What we have is... [a] system of organization in which corporations exercise the control and reap the benefits".*

Although the main concerns have been economic, the side effects of globalisation extend into all aspects of society. Globalisation also affects language and culture, following the same principles as economic globalisation. One symptom of cultural globalisation has been the spread of western culture and languages, especially English, which has led to the loss of native languages all over the world.

When considered globally, bioinvasions form a biological parallel to the processes that lead to cultural and economic globalisation (Emmeche 2001). Just as the spread of common languages, especially English, has led to a loss of cultural diversity, so too the globalisation of species via bioinvasion has contributed to local reductions in biological diversity in most parts of the world.

Invasions by exotic species transported by humans are second only to human land usage as a cause of recent extinctions. In both Australia and the United States, for example, over 1900 non-native plant species have become established and these constitute around 11 per cent of flora (Vitousek et al. 1997b). In some regions these figures are far higher: for example, introduced plants form around 40 per cent of the flora of the British Isles, New Zealand, Hawaii and Lord Howe Island, and over 60 per cent of that of Bermuda, Rodrigues, Ascension, and

Tristan da Cunha. Even small-seeming invasion events can cause massive ecological damage. For example, the introduction of a single foreign fish (the Nile perch) into a single African lake during the 1950s has since caused over 200 extinctions (Kolar and Lodge 2001).

Until recently, biological invasions were considered primarily as a problem for terrestrial ecology. Many terrestrial invaders are serious agricultural pests, giving governments an economic incentive to tightly monitor and regulate the movement of biological material. These controls have decreased the rate of invasions on land.

However, the same attention has not been given to marine ecosystems, where observation is more difficult and the consequences of introductions are less obvious. Even today, invasions of marine systems remain poorly documented, but there is increasing awareness that globalised shipping has had a devastating impact (Carlton and Geller 1993). As they travel around the world, ships dump and refill their ballast indiscriminately at different ports. Many marine organisms have tiny juvenile forms that can survive in the ballast water of ships. Chu et al. (1997) found at least 81 species in the ballast water of just five ships entering Hong Kong. As a result, certain coastal species have become ubiquitous worldwide, but the ecological results of these introductions remain largely uncounted (Carlton and Geller 1993). That ballast-water invasions can be disastrous is clear: for example, in 1991, 10,000 people died of cholera after its accidental introduction into Peru via the ballast water of a ship (Kolar and Lodge 2001).

Both invasions and extinctions contribute to a worldwide phenomenon that McKinney and Lockwood (1999) call *biotic homogenization*, in which a small number of 'winners' replace many 'losers'. The 'winners' in the current mass extinction are species that thrive in human-altered environments, such as house mice, rats, foxes, domestic pigeons, and a wide variety of weeds. Biotic homogenization is also noticeable in the mass extinctions of the past. McKinney and Lockwood note that the 'winners' in every case are not chosen at random, but share specific properties. They are ecological generalists and opportunists, adapted to disturbed environments. They tend to be omnivorous, fast-growing and widely dispersing, with high reproductive rates. Conversely, the species that are disappearing globally are the specialists, which lack these traits. The overall effect is that the world's biota are becoming more homogeneous, not only in terms of the number of species, but also in the ways in which they live.

Globalisation creates enormous problems, but it also offers opportunities in environmental management. Diamond (2000) commented that "Past societies faced frequent ecological crises of small amplitude over small areas. Modern global society faces less frequent but bigger crises over larger areas." The ready availability of global information allows us to look at problems and issues from a wider scale, and to benefit from the broad experience of different societies in different parts of the world.

Many past environmental problems have arisen because people acted to solve local problems in their immediate environment, without considering the broader consequences: we saw this in Chapter 1 in the cases of the Murray and the Amazon. The antidote to such local thinking is global information, and this is precisely the tool that globalisation offers.

11.4 THE CHANGING NATURE OF CONSERVATION

Some people have likened environmental information management to fiddling while Rome burns. Why build databases when entire ecosystems are being destroyed? The answer is that conservation management needs a sound basis in fact. We must base our solutions on knowledge and not be distracted by distorted perceptions generated by sensationalist media or seduced by narrowly-based, quick and dirty answers. To conserve populations successfully managers need to know what they are, where they are, and what factors and processes influence them. Information is essential also in setting conservation priorities. It is easy to spend a lot of effort saving a single rare species in the park next door while ignorant of hundreds that are endangered by clearing in remote areas. Information management therefore plays an essential role in conservation.

Compiling accurate information became an issue as conservation priorities shifted in many countries. For most of the 20th century, conservation could be equated with national parks. However, the rapidly growing scale of environmental alteration and increasing public awareness of environmental issues highlighted a need for off-reserve conservation and for broad-scale landscape management. The range of management issues is extensive. Examples include: environmental impact assessment, state of the environment reporting, environmental monitoring, conservation of rare and endangered species, natural heritage planning, species relocation programs, and land use planning.

One of the most important findings for conservation research is that most of the world's biodiversity is concentrated in a few biological 'hotspots'. In a famous study, Norman Myers and colleagues examined the global distributions of endemic species (Myers et al. 2000). They found that 44 per cent of vascular plants and 35 per cent of species in four vertebrate groups were confined to just 1.4 per cent of the earth's surface. This global view reveals the importance of looking beyond local, national and even continental scales in our efforts to conserve biodiversity. By targeting hotspots for protection, conservationists may achieve more effective preservation of biodiversity at a fraction of the cost.

Such protection is urgently needed. As Pimm and Raven (2000) commented,

"In the 17 tropical forest areas designated as biodiversity hotspots, only 12% of the original primary vegetation remains, compared with about 50% for tropical forests as a whole. Even within those hotspots, the areas richest in endemic plant species... have proportionately the least remaining vegetation and the smallest areas currently protected... Unless there is immediate action to salvage the remaining unprotected hotspot areas, the species losses will more than double."

Studies such as Myers et al. (2000) have contributed to a growing awareness that local decisions and priorities need to be set in a wider and ultimately global context. For instance to decide whether or not to log a patch of forest, planners need to know how much forest there is, what species might be put at risk, what the global costs and benefits are, and so on. Conversely, every local area contributes valuable data and experience that can be applied to other areas and can feed into global planning.

The informatics paradigm that has emerged treats environmental management as a host of conservation activities that reinforce each other, and are constrained by the global picture. Good communication is crucial. In terms of data sharing, this means making information widely available, with everyone contributing data. Access to relevant and reliable information is necessary to understand matters in context.. To this end, since the 1990s, governments have been very active in setting up regional, national, and international environmental information systems.

11.5   THE FUTURE

One of the biggest challenges for the future is human population growth. The species *Homo sapiens* has been around for only 200,000 years, or about 0.005 per cent of the time that life has existed on Earth. For most of that time, the global human population size was in the thousands. Indeed, it reached the billion mark (1,000,000,000 people) only in the late 1800s (Figure 5-1). By the time of writing, early in the new millennium, it had reached 6.2 billion. The implications of this exponential growth are enormous, and they are global by definition. The ecological lessons of the past are useful, but limited, because as a species we simply have never been in this position before.

Many people are shocked when confronted with the above reality. Of course it is not the number of people *per se* that is the problem, but the ecological footprint each person stamps on the environment. We consume, we exploit resources and we pollute and degrade ecosystems (Figure 11-1). Every new person is additional stress on these ecosystems, and there are currently 220,000 people being born every day.

So what are the lessons that we can learn from complexity? One important lesson is that we can no longer deal with issues in isolation. With globalisation now a well-recognised aspect of both economics and culture, the human race is rapidly approaching its limits to growth. Events no longer happen in isolation. Any major (or even minor) activity potentially impinges on people and places everywhere.

Some of these chains of interactions can be long and complex. Raise the price of soybeans and a chain of cause and effect might lead to vast tracts of rainforest being cast under threat. Try to save a unique wildflower and the flow-on effects could lead to a giant corporation going bankrupt.

Economic activity has always been based on the premise that growth is necessary; this is an ingrained assumption in economic theory. For centuries, western civilisation avoided the problems of achieving sustainable balance between exploitation and conservation by expanding their activities into the Third World. We saw earlier how in ancient times, many civilisations disappeared because they exhausted their environment. Western economies avoided this

Figure 11-1. Environmental degradation around Queenstown, Tasmania (photographed in 1984) arising from pollution associated with mining activity.

problem by expanding. They created empires from which they could extract materials that had become depleted at home. In more recent times, international corporations have extended that principle by moving manufacturing and production to Third World countries where they can exploit cheap labour and less stringent working conditions and relative freedom from environmental protection regulations.

But what will happen when we run out of Third World to exploit? Improvements in the efficiency of agricultural practices and the use of genetically enhanced crops are only stop-gap measures. The notion that science and technology will allow us to overcome the physical limits of the natural environment is fanciful. Sooner or later, growth has to halt. Sooner or later, world economics have to change. The possibility of steady-state economy has already been foreshadowed by many authors. Perhaps the most notable was E.A. Schumacher in his book *Small is Beautiful* (1989). Likewise, Hardin (1968) observed that we cannot simultaneously maximise human quality of life and continue exponential growth in a resource-limited world.

The central lesson from complexity theory is that we should be very wary of treating any environmental problem in isolation. We must continue to develop the infrastructure to deal with environmental issues in their geographical context, from the scale of local landscapes to global ecology. This infrastructure complements the traditional tools of local ecology with data warehouses, satellite imagery, network analysis, dynamical systems theory, geographic information systems, and simulation models. All of these new tools contribute to our ability to plan and test scenarios globally.

The manner of conservation has changed over the last decade. In the past, conservationists tended to focus on single species and single processes in pristine, 'natural' environments. The aim was to preserve fragments of untouched nature in national parks and other protected areas. Increasingly, we are aware that altered landscapes are the predominant reality in the world today, and that protected areas cannot be isolated from the surrounding environment. Altered landscapes cannot be ignored; they need to be an integral part of a world-wide conservation effort. With these changes in direction of conservation over the last decade has come an increased awareness by the general public of conservation issues. This has resulted in conservation taking more of a human perspective, with more public involvement.

In many traditional societies, some resources are shared among the community and collectively maintained by them. Such resources are called *commons*. The environment constitutes a form of global commons: a healthy environment benefits everyone, but effort is needed to maintain it. Models show that managing commons successfully is a difficult business. It is in the interests of every individual to cheat by ruthlessly exploiting the commons because if he does not do so now, sooner or later somebody else will. But when too many parties cheat, the commons collapses and everyone suffers. In a famous essay in *Science*, Hardin (1968) termed this seemingly inevitable disaster the 'tragedy of the commons'.

In traditional societies, commons are sustained by social pressure and well-defined rules limiting individual use. If the punishment for cheating is large enough, and people are vigilant in looking for cheats, then cheating is not worth the risk of being caught. However, applying this principle to environmental management is difficult. To define what constitutes cheating, we need a clear understanding of how the environment works. To detect cheating when it is happening, we need global monitoring systems. To punish cheats effectively, we need an environmental justice system. Given that the significant offenders are likely to be national governments or international corporations, these are very challenging requirements.

Ridley (1996) argues that clear private property rights are a key component of a successful commons. Self-restraint pays only when individual long-term rights to a resource are guaranteed. Supportive evidence for this idea comes from Maine lobster fisheries, which were apparently maintained sustainably for generations without any strong external regulation because of the existence of well-defined, though informal, territories. Likewise, Ridley claims that the decline of African wildlife caused by widespread hunting and poaching has been dramatically reversed whenever wildlife has been privatised. On this basis, Ridley argues that ideological opposition to private property is a major barrier to conservation.

The UNESCO Biosphere Reserve program is an attempt to develop the idea of an environmental commons at a local level by linking conservation with sustainable development (Fall 1999). Biosphere reserves are areas that contain a variety of zones managed with different levels of intensity, with the aim of conserving biodiversity while also benefiting the human inhabitants and other stakeholders. Likewise, world heritage areas, such as the Great Barrier Reef, are protected areas, and yet include human activity such as fishing and ecotourism within their management plans.

We began this book by describing why we need new ways of thinking, and new forms of data, to deal with complexity in landscapes. In subsequent chapters, we documented interacting processes at the levels of species, population and community, as well as the underlying role of genetics in the evolution of life on Earth. We have described how studies within 'natural' environments are becoming less appropriate for broader conservation, because altered landscapes are the dominant canvas on which we must work now and in the future. Instead of ignoring humans, or simply noting the species as the culprit, contemporary approaches to conservation must recognise human activities as landscape processes, include people as part of the solutions, and yield management plans that address the economic and social realities of human society.

This sounds complex. It would be tempting to break up these challenges into ever smaller bits that we could handle at local scales. The message of this book is that local actions have global consequences, and these consequences are often unexpected. If we are to maintain a habitable planet, the public, politicians, scientists and environmental managers need to accept the complexity of landscape ecology—with all its inherent challenges for conservation. This will require researchers to increasingly adopt the use of many of the tools described in this book, such as GIS and landscape imagery, global data warehouses and database mining, simulation and visualisation. Moreover, it will require the public and the leaders in our society to resist the assumption that there are simple answers to complex problems. It is imperative that environmental and social issues be addressed on geographical and temporal scales that are consistent with the challenges we face in our global village.

REFERENCES

Ackland, G.J. (2004). Maximization principles and Daisyworld. *J. Theor. Biol.* 227: 121–128.

Ackland, G.J., Clark, M.A. and Lenton, T.M. (2003). Catastrophic desert formation in Daisyworld. *J. Theor. Biol.* 223: 39–44.

Adami, C. (1994). *Learning and complexity in genetic auto-adaptive systems.* Technical Report. Pasadena, CA, California Institute of Technology.

Adami, C. and Brown, C.T. (1995). Evolutionary learning in the 2D artificial life system Avida. In R.A. Brooks and P. Maes, eds. *Artificial Life IV.* Cambridge, MIT Press. pp. 377–381.

Adami, C., Brown, C.T. and Haggerty, M.R. (1995). Abundance-distributions in the artificial life and stochastic models: Age and Area revisited. In F. Moran, A. Moreno, J.J. Merelo and P. Chacon, eds. *Advances in Artificial Life.* Berlin, Springer. pp. 503-514.

Alford, R.A. and Richards, S.J. (1999). Global amphibian declines: a problem in applied ecology. *Annu. Rev. Ecol. Syst.* 30: 133–65.

Allen, J.P. (2000). Artificial biospheres as a model for global ecology on planet Earth. *Life Support Biosph Sci.* 7: 273–82.

Allen, J.P, Nelson, M. and Alling, A. (2003). The legacy of biosphere 2 for the study of biospherics and closed ecological systems. *Adv. Space Res.* 31: 1629–1639.

Alvarez, L.W., Alvarez, W., Asaro, F. and Michel, H.V. (1980). Extraterrestrial cause for the Cretaceous-Tertiary extinction. *Science* 208: 1095–1108.

Armstrup, S.C. and Durner, G.M. (1995). Survival rates of radio collared female polar bears and their dependent young. *Can. J. Zool.* 73: 1312–1322.

Avilés, L. 1999. Cooperation and non-linear dynamics: an ecological perspective on the evolution of sociality. *Evol. Ecol. Res.* 1: 459–477.

Baillie, M. (1999). *Exodus to Arthur – Catastrophic Encounters with Comets.* London, Batsford.

Bak, P. and Chen, K. (1991). Self-organized criticality, *Sci. Am.* 265: 26–33.

Ball, M.C. (1988). Ecophysiology of mangroves. *Trees* 2: 129–142.

Ball, P. (1990). *The Self-made Tapestry: Pattern Formation in Nature.* Oxford, Oxford Univ. Press.

Barton, N.H. (1979). The dynamics of hybrid zones. Heredity 43: 341–359.

Beer, T. and Enting, I.G. (1990). Fire spread and percolation modelling. *Math. Comput. Model.* 13: 77–96.

Bennett, E.M., Carpenter, S.R. and Caraco, N.F. (2001). Human impact on erodable phosphorus and eutrophication: a global perspective. *BioScience,* 51: 227–234.

Bennett, K.D. (1983). Postglacial population expansion of forest trees in Norfolk, U.K. *Nature* 303: 164–167.

Bennett, K.D. (1985). The spread of Fagus grandifolia across North America during the last 18,000 years. *J. Biogeogr.* 12: 147–164.

Benton, M.J. and Twitchett, R.J. (2003). How to kill (almost) all life: the end–Permian extinction event. *Trends Ecol. Evol.* 18: 338–345.

Berglund, B.E. (2003). Human impact and climate changes – synchronous events and a causal link? *Quatern. Int.* 105: 7–12

Blackith, R. and Reyment, R.A. (1971). *Multivariate morphometrics.* New York, Academic Press.

Blaustein, A.R. and Kiesecker, J.M. (2002). Complexity in conservation: lessons from the global decline of amphibian populations. *Ecol. Lett.* 5: 597–608.

Boerlijst, M.C. and Hogeweg, P. (1991). Spiral wave structure in pre-biotic evolution: hypercycles stable against parasites. *Physica D* 48: 17–28.

Bolton, M.B. and Green, D.G. (1991). Computers and conservation – the Environmental Resources Information Network. *Trees and Natural Resources* 33: 14–16.

Bormann, F.H. and Likens, G.E. (1979). *Pattern and process in a forested watershed.* New York, Springer–Verlag.

Bossomaier, T.R.J. and Green, D.G. (1998). *Patterns in the Sand.* Sydney, Allen and Unwin.

Bossomaier, T.R.J. and Green, D.G. (2000). *Complex Systems.* New York, Cambridge Univ. Press.

Boston, T. and Stockwell, D.R. (1995). Interactive species distribution reporting, mapping and modelling using the World Wide Web. *Computer Networks and ISDN Systems* 28: 231–238.

Boustany, A.M., Davis, S.F., Pyle, P., Anderson, S.D., LeBoeuf, B.J. and Block, B.A. (2002). Expanded niche for white sharks. *Nature* 415: 35–36.

Bowman, D.M. (1998). The impact of Aboriginal landscape burning on the Australian biota. *New Phytologist* 140: 385–410.

Bradbury, R.H., van der Laan, J.D. and MacDonald, B. (1990). Modelling the effects of predation and dispersal on the generation of waves of starfish outbreaks. *Math. Comput. Model.* 13: 61–68.

Brauchli, K., Killingback, T. and Doebeli, M. (1999). Evolution of cooperation in spatially structured populations. *J. Theor. Biol.,* 200: 405–417.

Brewer, P.G. 2000. Gas hydrates: challenges for the future. *Ann. N. Y. Acad. Sci.* 912: 195–199.

Brown, J.H. and Lomolino, M.V. (2000). Concluding remarks: historical perspective and the future of island biogeography theory. *Global Ecol. Biogeogr. Lett.* 9: 87–92.

Bull, C.M. and Possingham, H. (1995). A model to explain ecological parapatry. *Am. Nat.* 145: 935–947.

Bulmer, M. (1989). Structural instability of models of sexual selection. *Theor. Popul. Biol.* 35: 195–206.

Burdet, H. M. (1992). What is IOPI? *Taxon* 41: 390–392.

Busby, J. (1991). BIOCLIM–a bioclimate analysis and prediction system. In C. Margules, M.P. Austin, eds. *Nature Conservation: Cost Effective Biological Surveys and Data Analysis.* Melbourne, CSIRO. pp. 64–67.

Cain, M.L., Andreasen, V. and Howard, D.J. (1999). Reinforcing selection is effective under a relatively broad set of conditions in a mosaic hybrid zone. *Evolution* 53: 1343–1353.

Caldarelli, G., Frondoni, R., Gabrielli, A., Montuori, M., Retzlaff, R. and Ricotta, C. (2001). Percolation in real wildfires. *Europhys. Lett.* 56: 510–516.

Carlton, J.T. and Geller, J.B. 1993. Ecological roulette: The global transport of non-indigenous marine organisms. *Science* 261: 78–82.

Chen, Y. (1988). Early Holocene population expansion of some rainforest trees at Lake Barrine basin, Queensland. *Aust. J. Ecol.* 13: 225–234.

Chu K.H., Tam, P.F., Fung, C.H. and Chen, Q.C. (1997). A biological survey of ballast water in container ships entering Hong Kong. *Hydrobiologia* 352: 201–206.

Chuck, A., Tyrrell, T., Toterdell, I.J. and Holligan, P.M. (2005). The oceanic response to carbon emissions over the next century: investigation using three ocean carbon cycle models. *Tellus B* 57: 70–86.

Clements, F.E. (1916). *Plant Succession.* Washington, Carnegie Inst. Publ. No. 242.

Cody, M.L. (1970). Chilean bird distribution. *Ecology* 51: 455–464.

Cohen, J.E. and Tilman, D. (1996). Biosphere 2 and biodiversity – the lessons so far. *Science* 274: 1150–1151.

Collins, J.J. and Chow, C.C. 1998. It's a small world. *Nature,* 393: 409–410.

Costantino, R.F., Desharnais, R.A., Cushing, J.M. and Dennis, B. (1997). Chaotic dynamics in an insect population. *Science* 275: 389–391.

Costanza, R. (1975). *The spatial distribution of land use subsystems, incoming energy and energy use in south Florida from 1900 to 1973.* Masters thesis, Florida, University of Florida.

Cox, P.M., Betts, R.A., Jones, C.D., Spall, S.A. and Totterdell, I.J. (2000). Acceleration of global warming due to carbon-cycle feedbacks in a coupled climate model. *Nature* 408: 184–187.

Crawley, M.J. (1990). The population dynamics of plants. *Philos. T. Roy. Soc. B* 330: 125–140.

D'Antonio, C.M. and Vitousek, P.M. (1992). Biological invasions by exotic grasses, the grass/fire cycle, and global change. *Annu. Rev. Ecol. Syst.* 23: 63–87.

Davies, P. (1998). Foreword. In Bossomaier, T.R.J. and Green, D.G. *Patterns in the Sand.* Allen and Unwin, Sydney, pp. vi-x.

Davis, M.B. (1976). Pleistocene biogeography of temperate deciduous forests. *Geoscience and Man* 13: 13–26.

Dawkins, R. 1976. *The Selfish Gene.* Oxford Univ. Press., New York.

Delcourt, P.A. and Delcourt, H.R. (1987). *Long–Term Forest Dynamics of the Temperate Zone.* Ecological Studies 63. New York, Springer-Verlag.

deMenocal, P.B. (2001). Cultural responses to climate change during the late Holocene. *Science* 292: 667–673.

Deutsch, C.J., Reid, J.P., Bonde, R.K., Easton, D.E., Kochman, H.I. and O'Shea, T.J. (2003). Seasonal movements, migratory behavior, and site fidelity of West

Indian manatees along the Atlantic Coast of the United States. *Wildlife Monogr.* 67: 1–77.

Diamond, J. (2000). Ecological collapses of pre-industrial societies. *The Tanner Lectures on Human Values.* Stanford University, May 22–24 2000. www.tannerlectures.utah.edu/lectures/Diamond_01.pdf

Dodge, M. (1999). Explorations in AlphaWorld: the geography of 3-D virtual worlds on the Internet. In *Virtual Reality in Geography – Workshop and Special Session at the RGS–IBG Annual Conference*, Leicester, 4–7th January 1999.

Duarte, J. (1997). Bushfire automata and their phase transitions. *Int. J. Mod. Phys. C* 8: 171–189.

Eigen, M. and Schuster, P. (1979). *The hypercycle: A principle of natural self–organization.* Berlin, Springer-Verlag.

Eldredge, N. and Gould, S.J. (1972). Punctuated equilibria: An alternative to phyletic gradualism. In T. J. M. Schopf, ed. *Models in Paleobiology.* Freeman, Cooper, San Francisco. pp. 82–115.

Elton, C.S. (1930). *Animal Ecology and Evolution.* New York, Oxford University Press.

Elton, C.S. (1958). *The Ecology of Invasions by Plants and Animals.* London, Methuen.

Emmeche, C. (2001). Bioinvasion, globalization, and the contingency of cultural and biological diversity – some ecosemiotic observations. *Sign Systems Studies* 29: 235–262.

Erdos, P., Renyi, A. (1960). On the evolution of random graphs, *Mat. Kutato. Int. Kozl.* 5: 17–61.

Erwin, D.H. (1998). The end and beginning: recoveries from mass extinctions, *Trends Ecol. Evol.* 13: 344–349.

Erwin, T.L. (1982). Tropical Forests: their richness in Coleoptera and other Arthropod Species, *Coleopts. Bull.* 36: 74–75

Erwin, T.L. (1988). The tropical forest canopy: the heart of biotic diversity. In E.O. Wilson, ed. *Biodiversity.* Washington, National Academy Press. pp. 123–9.

Fall, J.J. (1999). Transboundary biosphere reserves: a new framework for cooperation. *Environ. Conserv.* 26: 1–3.

Farkas, I., Derenyi, I., Jeong, H., Neda, Z., Oltvai, Z. N., Ravasz, E., Schubert, A., Barabasi, A-L. and Vicsek, T. 2002. Networks in life: scaling properties and eigenvalue spectra. *Physica A* 314: 25–34.

Fearnside, P.M. (1999). Biodiversity as an environmental service in Brazil's Amazonian forests: risks, value and conservation. *Environ. Conserv.* 26: 305–321.

Ferriere, R. and Gatto M. (1993). Chaotic population dynamics can result from natural selection. *Proc. Biol. Sci.* 251: 33–8.

Fisher, R.A. (1930). *The Genetical Theory of Natural Selection.* Oxford, Clarendon Press.

Fitz, H.C., DeBellevue, E.B., Costanza, R., Boumans, R., Maxwell, T., Wainger, L. and Sklar, F.H. (1996). Development of a general ecosystem model for a range of scales and ecosystems. *Ecol. Model.* 88: 263–295.

Flannery, T. (1994). *The Future Eaters.* Melbourne, Reed Books.

Flannery, T. (2001). *The Eternal Frontier: An Ecological History of North America and Its People.* New York, Atlantic Monthly.

Fox, B.J. and Fox, M.D. (2000). Factors determining mammal species richness on habitat islands and isolates: habitat diversity, disturbance, species interactions and guild assembly rules. *Global Ecol. Biogeogr. Lett.* 9: 19–37.

Frankham, R. (1998). Inbreeding and extinction: island populations. *Cons. Biol.* 12: 665.

Futuyma, D.J. and Moreno, G. (1988). The evolution of ecological specialization *Ann. Rev. Ecol. Syst.* 19: 207–33.

Gardner, M. (1970). Mathematical games: The fantastic combinations of John Conway's new solitaire game "life". *Sci. Am.* 223: 120–123.

Gardner, M. and Ashby, W.R. (1970). Connectance of large, dynamic cybernetic systems: critical values for stability. *Nature* 228: 784.

Gerlach, G. and Musolf, K. (2000). Fragmentation of landscape as a cause for genetic subdivision in bank voles. *Cons. Biol.* 14: 1066.

Gould, S.J. (2002). *The Structure of Evolutionary Theory.* Cambridge, Belknap Press.

Green, D.G. (1982). Fire and stability in the postglacial forests of southwest Nova Scotia, *J. Biogeogr.* 9: 29–40.

Green, D.G. (1983). Shapes of simulated fires in discrete fuels. *Ecol. Model.* 20: 21–32.

Green, D.G. (1987). Pollen evidence for the postglacial origins of Nova Scotia's forests. *Can. J. Bot.* 65: 1163–1179.

Green, D.G. (1989). Simulated effects of fire, dispersal and spatial pattern on competition within vegetation mosaics, *Vegetatio* 82: 139–153.

Green, D.G. (1990). Landscapes, cataclysms and population explosions. *Math. Comput. Model.* 13: 75–82.

Green, D.G. (1992). Ecology and conservation – the role of biological collections. *Australian Biologist* 5: 48–56.

Green, D.G. (1993). Emergent behaviour in biological systems. *Complexity International* 1. http://www.csu.edu.au/ci/vol1/David.Green/paper.html

Green, D.G. (1994a). Connectivity and complexity in ecological systems. *Pacific Conservation Biology* 1: 194–200.

Green, D.G. (1994b). Databasing diversity – a distributed, public–domain approach. *Taxon* 43: 51–62.

Green, D.G. (1994c). Connectivity and the evolution of biological systems. *J. Biol. Syst.* 2: 91–103.

Green, D.G. (2000). Self–organization in complex systems. In T.R.J. Bossomaier, and D.G. Green, eds. *Complex Systems.* New York, Cambridge University Press. pp. 7–41.

Green, D.G. (2004). *The Serendipity Machine.* Crows Nest, NSW, Allen and Unwin.

Green, D.G., and Klomp, N.I. (1997). Networking Australian biological research. *Australian Biologist* 10: 117–120.

Green, D.G. and Bossomaier, T.R.J. (2002). *Online GIS and Spatial Metadata*. London, Taylor and Francis.

Green, D.G. and Heng, T.N. (2005). *VLAB – The Artificial Life Virtual Laboratory*. Clayton, Auustralia, Monash University. http://www.complexity.org.au/vlab/

Green, D.G, Gill, A.M., and Tridgell, A. (1990). Interactive simulation of bushfire spread in heterogeneous fuel. *Math. Comput. Model.* 13: 57–66.

Green, D.G., Newth, D., Cornforth, D. and Kirley, M. (2001). On evolutionary processes in natural and artificial systems. In *Proceeding of the Fifth Australia- Japan Joint Workshop on Intelligent and Evolutionary Systems*. Dunedin, New Zealand, University of Otago. pp. 1–10.

Haken, H. (1981). *The Science of Structure: Synergetics*. New York, Van Nostrand Reinhold.

Hallam, A. and Wignall, P.B. (1997). *Mass Extinctions and their Aftermath*. Oxford, Oxford University Press.

Hanski, I. (1997). Metapopulation dynamics: From concepts and observations to predictive models. In I.A. Hanski, and M.E. Gilpin, eds. *Metapopulation Biology. Ecology, Genetics, and Evolution*. San Diego, Academic Press. pp. 69–91.

Hanski, I. (1999). An explosive laboratory. *Nature* 398: 387–388.

Hanski, I. and Gilpin, M. (1991). Metapopulation dynamics: brief history and conceptual domain. *Biol. J. Linn. Soc.* 42: 3–16.

Hardin, G. (1968). The tragedy of the commons. *Science* 162: 1243–1248.

Hardy, G. (1998). *OECD Megascience Forum Biodiversity Informatics Group*. http://www.oecd.org/ehs/icgb/BIODIV8.HTM

He, H.S. and Mladenoff, D.J. (1999). Spatially explicit and stochastic simulation of forest-landscape fire disturbance and succession. *Ecology* 80: 81–99.

Heaney, L.R. (2000). Dynamic disequilibrium: a long–term, large–scale perspective on the equilibrium model of island biogeography. *Global Ecol. Biogeogr. Lett.* 9: 59–74.

Held, I.M. and Soden, B.J. (2000). Water vapor feedback and global warming. *Annu. Rev. Energy. Environ.* 25: 441–475.

Herman, G.T. and Rozenberg. G. (1975). *Developmental Systems and Languages*. Amsterdam, North-Holland/American Elsevier.

Hogeweg, P. (1988). Cellular automata as a paradigm for ecological modeling, *Applied Math. Comput.* 27: 81–100.

Hogeweg, P. (1993). As large as life and twice as natural: bioinformatics and the artificial life paradigm. *Complexity International* 1: http://www.csu.edu.au/ci/vol1/

Hogeweg, P. and Hesper, B. (1983). The ontogeny of the interaction structure in bumblebee colonies: a MIRROR model. *Behav. Ecol. Sociobiol.* 12: 271–283.

Hogeweg, P. and Hesper, B. (1991). Evolution as pattern processing: TODO as a substrate for evolution. In J.A. Meyer and S.W. Wilson, eds. *From Animals to Animats*. Boston, MIT Press. pp. 492–497.

Holland, J. (1975). *Adaptation in Natural and Artificial Systems.* Ann Arbor, University of Michigan Press.

Holling, C.S. (1973). Resilience and stability of ecological systems. *Annu. Rev. Ecol. Sys.* 4: 1–23.

Hooper, D.S. (1999). *Cool School.* http://www.kewlschool.com/

House, S.M. (1985). *Relationship Between Breeding and Spatial Pattern in Some Dioecious Tropical Rainforest Trees.* PhD thesis, Canberra, Australian National University.

Huang, S., Pollack, H.N. and Shen, P.-Y. (2000). Temperature trends over the past five centuries reconstructed from borehole temperatures. *Nature* 403: 756–758.

Huxor, A. (1997). The role of virtual world design in collaborative working. *Proceedings–IEEE Conference on Information Visualisation (IV '97),* p. 246.

Ingersoll, A.P. (1969) The runaway greenhouse: A history of water on Venus. *J. Atmos. Sci.* 26: 1191–1198.

INPE (1998) *Deforestation in Brazilian Amazonia, 1995-1997,* Brazil, Instituto Nacional de Pesquisas Espaciais, e Ministério da Ciéncia e Tecnologia, Brasilia.

IUBS (International Union of Biological Sciences). (1998). *Species 2000.* www.sp2000.org/

Jeltsch, F. and Wissel, C. (1994). Modelling dieback phenomena in natural forests. *Ecol. Model.* 75: 111–121.

Karoly, D., Risbey, J. and Reynolds, A. (2003). *Global Warming Contributes To Australia's Worst Drought.* Sydney, WWF Australia.

Kasting, J.F. (1988). Runaway and moist greenhouse atmospheres and the evolution of Earth and Venus. *Icarus* 74: 472–94.

Kasting, J.F., Pavlov, A.A. and Siefert, J.L. (2001). A coupled ecosystem-climate model for predicting the methane concentration in the Archean atmosphere. *Origins Life Evol. B.* 31: 271–285.

Kauffman E.G. and Walliser O.H. (1990). Extinction Events in Earth History. *Lecture Notes in Earth Sciences.* Berlin, Springer-Verlag.

Kauffman, S.A. (1993). *The Origins of Order: Self-organization and Selection in Evolution.* Oxford, Oxford University Press.

Kaunzinger, C.M.K. and Morin, P.J. (1998). Productivity controls food-chain properties in microbial communities. *Nature* 395: 495–497.

Keddy, P.A. (1976). Lakes as islands: the distributional ecology of two aquatic plants, Lemna minor L. and L. trisulca L. *Ecology* 57: 163–359.

Kelly, M.J. and Durant, S.M. (2000). Viability of the Serengeti cheetah population. *Cons. Biol.,* 14: 786–797.

Kenrick, P. (1999). The family tree flowers. *Nature* 402: 358–359.

Kerr, J.T. (2001). Butterfly species richness patterns in Canada: energy, heterogeneity, and the potential consequences of climate change. *Conservation Ecology* 5: 10.

Kershaw, A.P., McKenzie, G.M. and McMinn, A. (1993). A Quaternary vegetation history of northeastern Queensland from pollen analysis of ODP site 820. In J.A. McKenzie, P.J. Davies and A. Palmer–Julson, eds.

Proceedings of the Ocean Drilling Program, Scientific Results 133, Houston, Texas A. and M. University. pp.107–114.

Keys, D. (1999). *Catastrophe — An Investigation into the Origins of the Modern World*. London, Arrow.

Kim, K.W. and Horel, A. (1998). Matriphagy in the spider *Amaurobius ferox* ([Araneae], Amaurobiidae): an example of mother-offspring interactions. *Ethology* 104: 1021–1037.

Kingsford, R.T. (2000). Ecological impacts of dams, water diversions and river management on floodplain wetlands in Australia. *Austral Ecology* 25: 109.

Kirchner, J.W. (2002). The Gaia hypothesis: fact, theory, and wishful thinking. *Climatic Change* 52: 391–408.

Kirkpatrick, J.B. and Bridle, K.L. (1999). Environment and floristics of ten Australian alpine vegetation formations. *Aust. J. Bot.* 47: 1–21.

Klomp, N.I. and Schultz, M.A. (2000). Short-tailed shearwaters breeding in eastern Australia forage in Antarctic waters. *Mar. Ecol. Prog. Ser.* 194: 307–310.

Klomp, N.I., Green, D.G. and Fry, G. (1997). Roles of technology in ecology. In N.I. Klomp, and L.D. Lunt, eds. *Frontiers in Ecology: Building the Links*. Oxford, Elsevier Science. pp. 299–309.

Knoll, A.H. (2003). *Life on a Young Planet: the First Three Billion Years of Evolution on Earth*. Princeton, Princeton University Press.

Knudtson, P. and Suzuki, D. (1992). *The Wisdom of the Elders*. Sydney, Allen and Unwin.

Koestler, A. (1967). *The Ghost in the Machine*. London, Hutchinson.

Kolar, C.S. and Lodge, D.M. 2001. Progress in invasion biology: predicting invaders. *Trends Ecol. Evol.* 16: 199–204.

Kriebel, D., Tickner, J., Epstein, P., Lemons, J., Levins, R., Loechler, E.L., Quinn, M., Rudel, R., Schettler, T. and Stoto, M. (2001). The precautionary principle in environmental science. *Environ. Health Perspect.* 109: 871–876.

Kuhn, T. (1962). *The Structure of Scientific Revolutions*. Chicago, University of Chicago Press.

Kutschera, U. and Wirtz, P. (2001). The evolution of parental care in freshwater leeches. *Theor. Biosci.* 120: 115–137.

Langton, C.G. (1990). Computation at the edge of chaos: phase transitions and emergent computation. *Physica D* 42: 12–37.

Langton, C.R., Burkhart, R. and Ropella, G. (1997). *The Swarm Simulation System*. Santa Fe, Santa Fe Institute. http://www.santafe.edu/projects/ swarm/

Lassila, O. and Swick, R.R. (1998). Resource Description Framework Model and Syntax. World Wide Web Consortium. http://www.w3.org/TR/1998/WD–rdf–syntax–19980216

Lawton, J.H. (1997). The role of species in ecosystems: aspects of ecological complexity and biological diversity. In T. Abe, S.A. Levin, M. Higashi, eds. *Biodiversity: An Ecological Perspective*. New York, Springer. pp. 215–228.

Lawton, J.H. and Brown, V.K. (1993). Redundancy in ecosystems. In E.D. Schulze, and H.A. Mooney, eds. *Biodiversity and Ecosystem Function*. Berlin, Springer. pp. 255–270.

Leakey, R., Lewin. R. (1994). *The Sixth Extinction – Patterns of Life and the Future of Mankind.* New York, Doubleday.

Lenton, T.M. and Lovelock, J.E. (2000). Daisyworld is Darwinian: Constraints on adaptation are important for planetary self-regulation. *J. Theor. Biol.* 206: 109–114.

Lenton, T.M. and van Oijen, M. (2002). Gaia as a complex adaptive system. *Philos. T. Roy. Soc. B* 357: 683–695.

Lett, C., Silber, C. and Barret, N. (1999). Comparison of a cellular automata network and an individual-based model for the simulation of forest dynamics. *Ecol. Model.* 121: 277–293.

Lever, C. (2001). *The Cane Toad. The History and Ecology of a Successful Colonist.* West Yorkshire, Westbury Academic and Scientific Publishing.

Levins, R. (1969). Some demographic and genetic consequences of environmental heterogenetiy for biological control. *Bulletin of the Entomological Society of America* 15: 237–240.

Levins, R. (1977). "The search for the macroscopic in ecosystems." In G.S. Innes, ed. *New Directions in the Analysis of Ecological Systems II.* La Jolla, Simulation Councils. pp. 213–222.

Lindenmayer, A. (1968). Mathematical models for cellular interaction in development. *J. Theoret. Biol.* 18: 280–315.

Lindenmayer, D.B. and Possingham, H.P. (1995). Modeling the viability of metapopulations of the endangered Leadbeater's possum in south–eastern Australia. *Biodivers. Conserv.* 4: 984–1018.

Lomolino, M.V. (2000). A call for a new paradigm of island biogeography. *Global Ecol. Bigeogr. Lett.* 9: 1–6.

London, S. (2000). Reassessing the Global Economy: An Interview with Jerry Mander. http://www.scottlondon.com/insight/scripts/mander.html

Lovelock, J.E. (1989). *The Ages of Gaia.* Oxford, Oxford Univ. Press.

Lovelock, J.E. and Margulis, L. (1974). Atmospheric homeostasis by and for the biosphere: the Gaia hypothesis. *Tellus* 26: 2–10.

Lynch, M., Conery, J. and Buerger, R. (1995). Mutation accumulation and the extinction of small populations. *Am. Nat.* 146: 489–518.

MacArthur, R.H. and Wilson, E.O. (1967). *The Theory of Island Biogeography.* Princeton, Princeton University Press.

Malamud, B.D. and Turcotte, D.L. (1999). Self-organized criticality applied to natural hazards. *Nat. Hazards* 20: 93–116.

Manrubia, S.C. and Solé, R.V. (1996). Self-organized criticality in rainforest dynamics. *Chaos Solitons and Fractals* 7: 523–541.

Manrubia, S.C. and Solé, R.V. (1997). On forest spatial dynamics with gap formation. *J. Theor. Biol.* 187: 159–164.

May, R.M. (1972). Will a large complex system be stable? *Nature* 238: 413–414.

May, R.M. (1973). *Stability and Complexity in Model Ecosystems.* Princeton, Princeton University Press.

May, R.M. (1976). Simple mathematical models with very complicated dynamics. *Nature* 26: 459–467.

May, R.M. (1988). How many species are there on Earth? *Science* 241: 1441–1449.

Mayr, E. (1942). *Systematics and the origin of species.* Columbia University Press, N.Y.

McKinney, M.L. and Lockwood, J.L. (1999). Biotic homogenization: a few winners replacing many losers in the next mass extinction. *Trends Ecol. Evol.* 14: 450–453.

McLure, J. (2003). A spatial assessment of landscape connectivity in the Apennine Mountains, Italy. *International Association for Landscape Ecology* (abstracts), Darwin, p. 87.

Merrilees, D. (1968). Man the destroyer: late Quaternary changes in the Australian marsupial fauna. *Journal of the Royal Society of Western Australia* 51: 1–24.

Metzger, J.P. (2003). Tree and vertebrate diversity according to forest area and connectivity in tropical forest fragments. *International Association for Landscape Ecology* (abstracts), Darwin.

Milgram, S. (1967). The small-world problem. *Psychol. Today,* 1: 60–67.

Millennium Ecosystem Assessment (2005). *Ecosystems and Human Well–Being: Desertification Synthesis.* Washington, World Resources Institute.

Miller, G.H., Magee, J.W., Johnson, B.J., Fogel, M.L., Spooner, N.A., McCulloch, M.T. and Ayliffe, L.K. (1999). Pleistocene extinction of *Genyornis newtoni*: human impact on Australian megafauna. *Science* 283: 205–208.

Miller, G.H., Fogel, M.L., Magee, J.W., Gagan, M.K., Clarke, S.J. and Johnson, B.J. (2005). Ecosystem collapse in Pleistocene Australia and a human role in megafaunal extinction. *Science* 309: 287–290.

Moilanen, A. and Hanski, I. (1998). Metapopulation dynamics: Effects of habitat quality and landscape structure. *Ecology* 79: 2503–2515.

Monsi, M. and Saeki, T. (1953). Über den Lictfaktor in den Pflanzengesell–schaften und sein Bedeutung fur die Stoffproduktion. *Jpn. J. Bot.* 14: 22–52.

Montoya, J.M. and Solé, R.V. (2002). Small world patterns in food webs. *J. Theor. Biol.* 214: 405–412.

Moore, P.D. (1999). Sprucing up beaver meadows. *Nature* 400: 622–621.

Moore, P.D. (1999). Woodpecker population drills. *Nature* 399: 528–529.

Myers, N., Mittermeier, R.A., Mittermeier, C.G., da Fonseca, G.A.B. and Kent, J. (2000). Biodiversity hotspots for conservation priorities. *Nature* 403: 853–858.

Naeem, S. (1999). Power behind diversity's throne. *Nature* 401: 653.

Nahmias, J., Téphany, H., Duarte, J. and Letaconnoux, S. (2000). Fire spreading experiments on heterogeneous fuel beds. Applications of percolation theory. *Can. J. Forest Res.* 30: 1318–1328.

Nakamaru, M., Matsuda, H. and Iwasa, Y. (1997). The evolution of cooperation in a lattice-structured population. *J. Theor. Biol.* 184: 65–81.

Nakamaru, M., Nogami, H. and Iwasa, Y. (1998). Score-dependent fertility model for the evolution of cooperation in a lattice. *J. Theor. Biol.* 194: 101–124.

Newman, M.E.J. 1997. A model of mass extinction. *J. Theor. Biol.* 189: 235–252.

Newth, D., Lawrence, J. and Green, D.G., (2002) Emergent organization in Dynamic Networks. In A. Namatame, D. Green, Y. Aruka and H. Sato, eds. *Complex Systems 2002*. Tokyo, Chuo University. pp. 229–237.

Noble, I.R. and Slatyer, R.O. (1981). Concepts and models of succession in vascular plant communities subject to recurrent fire. In A.M. Gill, R.H. Groves, and I.R. Noble, eds. *Fire and the Australian Biota*. Canberra, Australian Academy of Science. pp. 311–335.

Nordhaus, W.D. (2001). Climate change: global warming economics. *Science* 294: 1283–1284.

Nowak, M.A. and May, R.M. (1992). Evolutionary games and spatial chaos. *Nature* 359: 826–829.

NTA (2002). *Endangered Places – 2002 Report Card*. Canberra, National Trust of Australia.

Oblinger, D.G. and Oblinger, J.L., eds. (2005). *Educating the Net Generation*. www.educause.edu/educatingthenetgen/

O'Brien, S.J., Roelke, M.E., Marker, L., Newman, A., Winkler, C.A., Meltzer, D., Colly, L., Evermann, J.F., Bush, M. and Wildt, D.E. (1985). Genetic basis for species vulnerability in the cheetah. *Science,* 227:1428–34.

Odum, H.T. (1996). Scales of ecological engineering. *Ecol. Eng.* 6: 7– 19.

Overpeck, J.T. (2000). The hole record. *Nature* 403: 714–715.

Paine, R.T. (1966). Food web complexity and species diversity. *Am. Nat.* 100: 65–75.

Papert, S. (1973). *Uses of Technology to Enhance Education*. LOGO Memo no. 8, Boston, M.I.T. Artificial Intelligence Laboratory.

Pech, R., McIlroy, J.C., Clough, M.F. and Green, D.G. (1992). A microcomputer model for predicting the spread and control of foot and mouth disease in feral pigs. In J.E. Borrecco and R.E. Marsh, eds. *Proceedings of the 15th Vertebrate Pest Conference*. Davis, University of California. pp. 360–364.

Pepper, J.W. (2000). An agent-based model of group selection. In C.C. Maley and E. Boudreau, eds. *Artificial Life 7 Workshop Proceedings*. Cambridge, MIT Press. pp. 100–103.

Phillips, J.D. (1993). Biophysical feedbacks and the risks of desertification. *Ann. Assoc. Am. Geogr.* 83: 630–640.

Pielou, E.C. (1969). *An Introduction to Mathematical Ecology*. New York, Wiley Interscience.

Pielou, E.C. (1974). *Population and Community Ecology*. New York, Gordon and Breach.

Pielou, E.C. (1975). *Ecological Diversity*. New York, Wiley Interscience.

Pimm, S.L. (1982). *Food Webs*. London, Chapman and Hall.

Pimm, S.L. (1984). The complexity and stability of ecosystems. *Nature* 307: 321–326.

Pimm, S.L. and Raven, P. (2000). Extinction by numbers. *Nature* 403: 843–845.

Ponting, C. (1991). *A Green History of the World*. London, Penguin.

Poon, C. and Barahona, M. (2001). Titration of chaos with added noise. *P. Natl. Acad. Sci. USA*. 98: 7107–7112.

Popper, K.R. (1968). *The Logic of Scientific Discovery*. London, Hutchinson.

Poskanzer, J. (1991). *Xantfarm – simple ant farm for X11*. http://www.acme.com/software/xantfarm/

Post, D.M., Pace, M.L. and Hairston, N.G. (2000). Ecosystem size determines food-chain length in lakes. *Nature* 405: 1047–1049.

Prigogine, I. (1980). *From Being to Becoming*. San Francisco, W. H. Freeman and Co.

Prusinkiewicz, P. and Lindenmayer, A. (1990). *The Algorithmic Beauty of Plants*. Berlin, Springer–Verlag.

Puigdefabregas, J., Gallart, F., Biaciotto, O., Allogia, M., and del Barrio, G. (1999). Banded vegetation patterning in a subantarctic forest of Tierra del Fuego, as an outcome of the interaction between wind and tree growth. *Acta Oecol.* 20: 135–146.

Quinlan, R. (1986). Induction of decision trees. *Mach. Learn.* 1: 81–106.

Raup, D.M. (1986). Biological extinction in Earth history. *Science* 231: 1528.

Raup, D.M. and Jablonski, D. (1986). *Patterns and Processes in the History of Life*. Berlin, Springer-Verlag.

Ray, T. (1991). An approach to the synthesis of life. In C.G. Langton, C. Taylor, J.D. Farmer, S. Rasmussen, eds. *Artificial Life II*. New York, Addison-Wesley. pp. 41–91.

Raymond, R. (1986). *Starfish Wars – Coral Death and the Crown-of-Thorns*. South Melbourne, Macmillan Australia.

Recher, H.F. (1969). Bird species diversity and habitat diversity in Australia and North America. *Am. Nat.* 103: 75–80.

Rees, M., Kelly, D. and Bjoernstad, O. N. (2002). Snow tussocks, chaos and the evolution of mast seeding. *Am. Nat.* 160: 44–59.

Reynolds, C.W. (1987). Flocks, herds, and schools: a distributed behavioral model. *Computer Graphics* 21: 25–34.

Richardson, B.J. and McKenzie, A.M. (1992). Australia's biological collections and those who use them. *Australian Biologist* 5: 19–30.

Richmond, B. (1993). Systems thinking: critical thinking skills for the 1990s and beyond. *Syst. Dynam. Rev.* 9: 113–133.

Ridley, M. (1996). *The Origins of Virtue*. Viking, London.

Romme, W.H., Turner, M.G., Wallace, L.L. and Walker, J.S. (1997). Aspen, elk, and fire in northern Yellowstone National Park. *Ecology* 76: 2097–2106.

Roshier, D.A., Robertson, A.I., Kingsford, R.T. and Green, D.G. (2001). Continental–scale interactions with temporary resources may explain the paradox of large populations of desert waterbirds in Australia. *Landscape Ecol.* 16: 547–556.

Sadedin, S. and Littlejohn, M.J. (2003). A spatially explicit individual-based model of reinforcement in hybrid zones. *Evolution* 57: 962–970.

Sankaran, M. and McNaughton, S.J. (1999). Determinants of biodiversity regulate compositional stability of communities. *Nature* 401: 691–693.

Sato, K. and Iwasa, Y. (1993). Modelling of wave regeneration in subalpine Abies forests–population dynamics with spatial structure. *Ecology* 74: 1538–1550.

Satulovsky, J.E. (1997). On the synchronizing mechanism of a class of cellular automata. *Physica A* 237: 52–58.

Schaffer, M. (1987). Minimum viable populations: coping with uncertainty. In M.E. Soule, ed. *Viable Populations for Conservation*. Cambridge, Cambridge University Press. pp. 69–86.

Scholin, C.A., Gulland, F., Doucette, G.J., Benson, S., Busman, M., Chavez, F.P., Cordaro, J., Delong, R., De Vogelaere, A., Harvey, J., Haulena, M., Lefebvre, K., Lipscomb, T., Loscutoff, S., Lowenstine, L.J., Marin III, R., Miller, P.E., McLellan, W.A., Moeller, P.D.R., Powell, C.L., Rowles, T., Silvagni, P., Silver, M., Spraker, T., Trainer, V. and Van Dolah, F.M. (2000). Mortality of sea lions along the central California coast linked to a toxic diatom bloom. *Nature* 403: 80–84.

Schumacher, E.F. (1989). *Small is Beautiful: Economics as if people mattered*. New York, Harper Perennial.

Sharp, N.C. (1997). Timed running speed of a cheetah (*Acinonyx jubatus*). *J. Zool.*, 241: 493–494.

Siegenthaler, U. and Sarmiento, J.L. (1993). Atmospheric carbon dioxide and the ocean. *Nature* 365: 119–125.

Simberloff, D., Farr, J.A., Cox, J. and Mehlman, D.W. (1992). Movement corridors: conservation bargains or poor investments? *Cons. Biol.* 6: 493–504.

Singh, G. and Geissler, E.A. (1985). Late Cainozoic history of vegetation, fire, lake levels and climate, at Lake George, New South Wales, Australia. *Philos. T. Roy. Soc. B* 311: 379–447.

Solé, R.V. and Manrubia, S.C. (1996). Extinction and self-organized criticality in a model of large-scale evolution. *Phys. Rev. E* 54: R42–R45.

Solé, R.V., Manrubia, S.C., Benton, M. and Bak, P. (1997). Self-similarity of extinction statistics in the fossil record. *Nature* 388: 764–767.

Solé, R.V. and Montoya, J.M. (2001). Complexity and fragility in ecological networks. *Proc. R. Soc. Ser. B-Bio.* 268: 1–7.

Staley, M. (2002). Darwinian Selection Leads to Gaia. *J. Theor. Biol.* 218: 35–46.

Stauffer, D. (1979). Percolation. *Physics Reports* 54: 1–74.

Stockwell, D.R. (1992). *Machine Learning and the Problem of Prediction and Explanation in Ecological Modelling*. PhD thesis, Canberra, Australian National University.

Stockwell, D.R.B., Davey, S.M., Davis, J.R. and Noble, I.R. (1990). Using decision trees to predict Greater Glider density. *AI Applications* 4: 33–43.

Stott P.A., Stone, D.A. and Allen, M.R. (2004). Human contribution to the European heatwave of 2003. *Nature* 432: 610–14.

Suarez, A.V., Tsutsui, N.D., Holway, D.A. and Case1, T.J. (1999). Behavioral and genetic differentiation between native and introduced populations of the Argentine Ant. *Biological Invasions* 1: 43–53.

Sugimoto, T. (2002). Darwinian evolution does not rule out the Gaia hypothesis. *J. Theor. Biol.* 218: 447–455

Sunstein, C.R. (2003). Beyond the precautionary principle. *John M. Olin Law and Economics Working Paper No. 149*. Chicago, University of Chicago.

Sutherst, R. and Maywald, G.F. (1985). A computerised system for matching climates in ecology. *Agr. Ecosyst. Environ.* 13: 281–299.

Tangley, L. (1998). How many species exist? *Natl. Wildlife* 37: 32–33.

Terwilliger, J. and Pastor, J. (1999). Small mammals, ectomycorrhizae, and conifer succession in beaver meadows. *Oikos* 85: 83–94.

Thomas, C.D., Cameron, A., Green, R.E., Bakkenes, M., Beaumont, L.J., Collingham, Y.C., Erasmus, B.F.N., de Siqueira, M.F., Grainger, A., Hannah, L., Hughes, L., Huntley, B., van Jaarsveld, A.S., Midgley, G.F., Lera Miles, L., Ortega–Huerta, M.A., Peterson, A.T. Phillips, O.L. and Williams, S.E. (2004). Extinction risk from climate change. *Nature* 427: 145–148.

Thornton, I.W.B. (1996). *Krakatau. The destruction and reassembly of an island ecosystem*. Cambridge, Harvard University Press.

Tilman, D. (1996). Biodiversity: Population versus ecosystem stability. *Ecology* 77: 350–363.

Tregonning, K. and Roberts A. (1979). Complex systems which evolve towards homeostasis. *Nature* 281: 563–564.

Tsukada, M. (1982). *Cryptomeria japonica*: glacial refugia and late-glacial and post glacial migration. *Ecology* 63: 1091–1105.

Tsutsui, N.D. and Case, T.J. (2001). Population genetics and colony structure of the argentine ant (*Linepithema humile*) in its native and introduced ranges. *Evolution* 55: 976–985.

Turcotte, D.L. (1999). Applications of statistical mechanics to natural hazards and landforms. *Physica A* 274: 294–299.

Turner, M.G. (1989). Landscape ecology: the effect of pattern on process. *Annu. Rev. Ecol. Syst.* 20: 171–197.

Turner, M.G., Dale, V.H. and Everham III, E.H. (1997a). Fires, hurricanse and volcanoes. *BioScience* 47: 758–768.

Turner, M.G., Romme, W.H., Gardner, R.H. and Hargrove, W.W. (1997b). Effects of fire size and pattern on early succession in Yellowstone National Park. *Ecol. Monogr.* 67: 411–433.

UIUC Imaging Laboratory. (1998). *SmartForest – an interactive forest visualizer*. http://www.imlab.uiuc.edu/smartforest/

Usher, M.B. (1987). Effects of fragmentation on communities and populations: a review with applications to wildlife conservation. In *Nature Conservation: the Role of Remnants of Native Vegetation*. D.A. Saunders, G.W. Arnold, A.A. Burbidge, A.J.M. Hopkins, eds. Australia, Surrey Beatty and Sons. pp. 103–121.

Van der Laan, J.D. and Hogeweg, P (1992). Waves of crown–of–thorns starfish outbreaks–where do they come from? *Coral Reefs* 11: 207–213.

Van der Ree, R. (2003). *Ecology of Arboreal Marsupials in a Network of Remnant Linear Habitats*. PhD Thesis. Deakin University, Melbourne.

Vinnikov, K.Y. and Grody, N.C. (2003). Global warming trend of mean tropospheric temperature observed by satellites. *Science* 302: 269–272.

Vitousek, P.M., Aber, J.D., Howarth, R.W., Likens, G.E., Matson, P.A., Schindler, D.W., Schlesinger, W.H. and Tilman. D.G. (1997a). Human alteration of the global nitrogen cycle: sources and consequences. *Ecol. Appl.* 7: 737–750.

Vitousek, P.M., D'Antonio, C. M., Loope, L. L., Rejmánek, M. and Westbrooks, R. (1997b). Introduced species: a significant component of human–caused global change. *New Zeal. J. Ecol.* 21: 1–16.

Vitousek, P.M., Ehrlich, P.R., Ehrlich, A.H. and Matson, P.A. (1986). Human appropriation of the products of photosynthesis. *Bioscience* 36: 368–373

von Bertalanffy, L. (1968). *General Systems Theory: Foundations, Development, Applications*. New York, Braziller.

W3C (2005). *Extensible Markup Language (XML)*. http://www.w3.org/XML/

Wackernagel, M., Schulz, N.B., Deumling, D., Linares, A.C., Jenkins, M., Kapos, V., Monfreda, C., Loh, J., Myers, N., Norgaard, R. and Randers, J. (2002). Tracking the ecological overshoot of the human economy. *P. Natl. Acad. Sci. USA* 99: 9266–9271.

Walsworth, N.A. and King, D.J. (1999). Image modelling of forest changes associated with acid mine drainage. *Comput. Geosci.* 25: 567–580.

Walther, G–R., Post, E., Convey, P., Menzel, A., Parmesank, C., Beebee, T.J.C., Fromentin, J.M., Hoegh–Guldberg, O. and Bairlein, F. (2002). Ecological responses to recent climate change. *Nature* 416: 389–395.

Watts, D.J. and Strogatz, S.H. (1998). Collective dynamics of 'small–world' networks. *Nature* 393: 440–442.

Webb, T. III (1979). The past 11,000 years of vegetational change in eastern North America, *BioScience* 31: 501–506.

Weibel, S., Kunze, J. and Lagoze C. (1998) Dublin core metadata for simple resource discovery. *Dublin Core Workshop Series*. http://dublincore.org/

Weigle, B.L., Wright, I.E., Ross, M. and Flamm, R. 2001. Movements of radio–tagged manatees in Tamap Bay and along Florida's west coast, 1991–1996. *FMRI Technical Reports* TR–7, St. Petersburg, Florida, FMRI.

White, J.P. (1994). Site 820 and the evidence for early occupation in Australia. *Quaternary Australasia* 12: 21–23.

Whittaker, R.J. (1998). *Island biogeography: Ecology, evolution and conservation*. Oxford, Oxford University Press.

Whittaker, R.J. (2000). Scale, succession and complexity in island biogeography: are we asking the right questions? *Global Ecol. Biogeogr. Lett.* 9: 75–85.

Wiechert, U.H. (2002). Earth's early atmosphere. *Science* 298: 2341–2342.

Wilke, C.O. and Adami, C. (2002). The biology of digital organisms *Trends Ecol. Evol.* 17: 528–532.

Wilkinson, D. and Willemsen, J.F. (1988). Invasion percolation: a new form of percolation theory. *J. Phys. A – Math. Gen.* 16: 3365–3376.

Wilkinson, G.S. (1984). Reciprocal food sharing in the vampire bat. *Nature* 308: 181–184.

Williamson, M. (1996). *Biological Invasions. Population and Community Biology Series, Vol 15*. London, Chapman and Hall. p. 104, Figure 4.9.

Wilson, E.O. (ed.) (1988). *Biodiversity*. Washington, National Academy Press.

Wilson, E.O. (1992). *The Diversity of Life*. London, Penguin.

Winchester, S. (2003). *Krakatoa; The Day the World Exploded: August 27, 1883*. New York, Harper-Collins.

Wolfram S. (1984). Cellular automata as models of complexity. *Nature* 311: 419–424.

Wolfram S. (1986). *Theory and Applications of Cellular Automata*. Singapore, World Scientific.

Woodward, F.I. (1993). How many species are required for a functional ecosystem? In E.D. Schulze, and H.A. Mooney, eds. *Biodiversity and ecosystem function*. Berlin, Springer. pp. 271–291.

Wright, R. (1986). How old is Zone F at Lake George? *Archaeology in Oceania* 21: 138–139.

Würsig, B.F., Cipriano, F. and Würsig, M. (1991). Dolphin movement patterns: information from radio and theodolite tracking studies. In K. Pryor and K. S. Norris, eds. *Dolphin societies: Methods of study*. Berkeley, CA, University of California Press. pp. 79–111.

INDEX

abiotic, 7, 10, 92, 93, 121

acidification, 176, 177

adaptation, 116, 118-124, 126, 128, 130-132

aerial photograph, 34, 36, 42

Africa, 175, 182, 186

agent, 12, 24, 27-31, 36, 38, 63, 128, 131, 141, 142, 148, 164

algorithm, 13, 17, 30, 108, 142, 157

Amazon, 2, 3, 5, 9, 11, 171, 174, 182

analogy, 42, 54, 64, 106

ant, 4, 14, 29, 30, 32, 52, 63, 116, 128, 142

ant colony, 4, 29, 32, 52, 63

ant sort, 29, 30, 142

Antarctic, 38, 153, 178, 180

Arctic, 9, 71, 154

atomic fission, 49

attractor, 76, 77, 79, 81, 82, 115, 144

attribute, 17, 18, 24, 28, 34, 35, 56, 90, 105, 106, 119, 165, 180

Australia, 1, 5, 6, 33, 39, 48, 49, 69, 86, 89, 101, 106, 116, 121, 132, 140, 153, 155, 159, 163, 166, 167, 169, 173-176, 179-181

automaton, 30, 36, 37, 38, 40, 47, 90, 93, 101, 118, 119, 123, 125, 139, 143, 145

balance of nature, 68, 69, 70, 72, 74, 76, 78, 80, 82, 84, 85

bear, 50, 153

biochemical, 17, 23, 24, 26, 62

biochemical gradient, 17, 24

biodiversity, 2, 6, 6, 53, 111, 155, 162, 164, 165, 167-169, 173, 174, 183, 186

biological pattern, 44

biosphere, 10, 11, 132, 146, 176, 186

biotechnology, 13, 17

boid, 29-32, 142

Britain, 41, 43, 97, 99

CA (cellular automaton), 13, 30, 35-38, 40, 41, 47, 60, 90-93, 101, 102, 118, 121, 123, 125, 139, 141-146

calibration, 140

California, 69, 108, 109, 153

Canada, 39, 69, 95, 97

canopy, 2, 16, 17, 22, 23, 83, 84, 90, 159

cascade, 3

cell, 10, 18, 19, 23, 35, 36, 37, 40, 41, 44-49, 52, 53, 58, 60, 62, 65, 90, 117-119, 123, 125, 140-145, 160

cellular automaton, 30, 36-38, 40, 47, 90, 93, 101, 118, 119, 123, 125, 139, 143, 145

chaos, 8, 44, 66, 72, 78-82

closed system, 63, 64

clump formation, 58

colonist, 22, 90, 116, 128, 173

colonisation, 47, 69, 83, 84, 87, 90, 173, 174

colony, 4, 29, 32, 52, 63, 101

colony,ant, 4, 29, 32, 52, 63

communication, 12, 23, 54, 65, 127, 165, 180, 184

community, ecological, 106-112, 115, 150, 155, 160, 161, 180, 181, 185, 186

complex system, 5, 8, 13, 14, 53, 55,